In the Beginning: A Look at our World through the Lens of Environmental Science and Christian Theology

In the Beginning: A Look at our World through the Lens of Environmental Science and Christian Theology

Christopher S. Farabaugh PhD.
and
Timothy M. Farabaugh M.Div.

In the Beginning: A Look at our World through the Lens of Environmental Science and Christian Theology

© Christopher S. Farabaugh PhD. And Timothy M. Farabaugh M. Div. 2018

All rights reserved. Without limiting the rights under copyright reserved above, no part of this publication may be reproduced, stored in a retrieval system, or transmitted, in any form or by any means (electronic, mechanical, photocopying, recording or otherwise), without the prior written permission of the copyright owner of this book.

Published by
Lighthouse Christian Publishing
SAN 257-4330
5531 Dufferin Drive
Savage, Minnesota, 55378
United States of America

www.lighthousechristianpublishing.com

We are going to begin this book by looking at a few examples of creation myths from ancient cultures. These stories vary greatly. Some have good gods verses evil ones. Others have gods that fight and devour each other. Most of them explain, in some way, how men and women and the rest of creation came to be.

After we have introduced these examples, in the second section of the book, we will look at other theories of creation and evolution. We will follow that with a detailed look at the first creation story found in the Old Testament of the Bible. As we look at the creation story a day at a time, we will introduce scientific information regarding what had been created that day, per the biblical story.

The final section of the book will encourage the reader to look at how we can preserve that which has been created.

We are hopeful that the information as well as the questions at the end of each chapter will inspire you to think about what we have presented and encourage you to consider how amazing this thing is we call creation.

Section One

Myths

We begin this section with some Greek mythology. This information comes to us from Edith Hamilton who wrote the book entitled *Mythology*. As the story goes, there was only the formless confusion of Chaos that brooded over unbroken darkness. Two children were born to this shapeless nothingness. They were Night and Erebus, which is the unfathomable depth where death dwells.

From darkness and death Love is born, and with its birth, order and beauty began to languish behind confusion. Love created Light, and with its compassion, radiant Day.

Earth was the solid ground yet vaguely a personality, too. Heaven was the blue vault on high, but it acted in some ways, as a human being would.

The first creation who had the appearance of life were the children of mother Earth and Father Heaven (Gaea and Ouramos). The Greeks had no life form known to man.

Eventually, there were three types of creatures created. Three of them were monstrously huge and strong and each had a hundred hands and fifty heads. To three others was given the name Cyclopes because each had only one enormous eye. Last came the Titans. There were a large number of these and they were in no way inferior to others in size and strength, but they were not purely destructive.

Cronus was the Lord of the universe with his sister Rhea. One of the sons, the future ruler of the heavens and the earth, whose name in Greek is Zeus or in Latin, Jupiter, rebelled against him. Zeus won the battles. The victory of the radiant powers of Heaven over the brutal forces of Earth was complete. From then on Zeus and his brothers and sisters ruled, undisputed lords of all.

Humans had not been created yet, but the world, now cleared of the monsters, was ready for mankind. The earth was believed to be a round disk divided into two equal parts by the Sea, as the Greeks called it, which we know as the Mediterranean, and by what we call the Black Sea. [1]

The second set of creation stories are from a book entitled, *In the Beginning*, by Virginia Hamilton and Barry Moser. They say, in the introduction to the book, that the most striking purpose of a creation myth is to explain something. Yet it also asks questions and gives reasons why groups of people perform certain rituals and live in a particular way.

The first myth they share in this book is that of the "Raven as the Creator". It is an Eskimo myth that has man born of a pea pod the Raven had created. When the Raven saw the full-grown man, he continued to create other animals and finally a woman to be a helper and mate to the first man.[2]

The second myth from this set of creation stories is the account of "Quat the Creator". It is about Quat the solar god of the Banks Islands, north of the New Hebrides in Melanesia. According to this myth, in the beginning there was light. It never dimmed, this light covered everything. It was brightness that shown everywhere. Under the light was a huge stone. The stone was the mother, Quatgoro. Quatgoro split in half, and there came twelve sons into the light. They were Quat and his eleven brothers.

---

[1] Edith Hamilton, *Mythology*, (New York, Boston & London), 1942, pp. 67-83.
[2] Virginia Hamilton and Barry Moser, *In the Beginning*, (San Diego, New York, London), 1988. pp. 3-7.

The brothers were all named Tangaro, but they were not the same. The first brother after Quat was Tangaro the Wise. The second was Tangaro the Fool. The names of the other nine were names of leaves such as Breadfruit Leaf, Coconut Leaf, Bamboo Leaf, and so on. They all grew up as soon as they were born, just as Quat had.

Quat named himself when he was born since he had no father to name him. Right away he began to think about how he could make men. He also thought about how he could make plants, pigs and stones.

Quat made the first human from a tree. He carved arms and then legs, and he made the rest of the body apart from them. He made fingers and hands, toes and feet. He made ears and eyes-all neatly and carefully. Then he fitted the parts together. H made six of the wooden puppets.

And he stood them in a line, so he could do a sacred dance in front of them. Soon, the puppets began to move. They moved just a little at first; they moved stiffly. But they moved. Quat beat on his sacred drum. The drumbeats were like magic. The puppets moved faster until they were doing the dance of life to the drumbeat.

Now, the puppets that had life could stand, walk, and run along. Quat fixed them into men and women. There were three women and three men. Each of the women had a husband and each of the men had a wife.

After Quat made men and women, he decided to make pigs. At first, he had them stand up on two legs and walk that way. When his brothers saw this, they pointed and laughed, saying, "Your pigs look like men!" Quat didn't want the pigs to be laughed at, so, he shortened the front legs. Now the pegs walked on four feet instead of two.

In this way Quat made many things. He thought to make all kinds of plants, canoes and much more.[3]

The third myth found in this book by Hamilton and Moser is entitled, "An Endless Sea of Mud." It is a creation story from the Kono people of Guinea. Per the story, in the beginning, there was darkness, and in it lived Death, called Sa, and his wife and daughter. The three of them were all that was.

There was nowhere for them to live comfortably, so Sa started it. He used his magic power, and he made an endless mud sea. In this mud sea, Sa built his house.

After that, the god, Alatangana came to visit Sa. He found Sa's house dirty and dark. Alatangana thought Sa should do better than that, and he said so.

"Nothing can live in such a place," the god told Sa. "This house needs fixing up. Everything is too dark."[4]

So, Alatangana took things into his own hands. He made the mud solid. We know it as Earth. "The Earth feels sad" god said. "I will make plants and vegetables and animals to live on it." And so, he did.[5]

"Bursting from the Hen's Egg" is the name of the fourth creation myth in this collection. It comes from China and dates to 600 B.C. According to this myth, the shape of the universe was in the shape of a hen's egg. Within the egg was a great mass called nothing. Inside nothing was something not yet born. It was not yet developed, and it as called Phan Ku.

In no time, Phan Ku burst from the egg. He was the first being. He was the great Creator. Phan Ku was the size of a giant. He grew ten feet a day and lived for eighteen thousand years.

---

[3] *Ibid.* pp. 9-13
[4] *Ibid* p. 15.
[5] *Ibid* p.15.

Hair grew all over Phan Ku. Horns curved up out of his head, and tusks jutted from his jaw. In one hand, he held a chisel and with it he carved out the world.

Phan Ku separated sky from earth. The light, pure sky was Yang, and the heavy, dark weight of earth was Yin. The vast Phan himself filled the space earth and sky, Yin and Yang.

He chiseled out earth's rivers; he scooped out the valley's. It was easy for him to layer the mountains.

Then Phan Ku placed the stars and the moon in the night sky and the sun by day. He put the great seas where they are now, and he showed the people how to build ships and how to build bridges.

Only when Phan Ku died was the world complete at last. The dome of the sky was made of Phan Ku's skull. Solid soil was formed from his bones, rivers and seas from his blood. All of plant life came from Phan Ku's hair. Thunder and lightning are the sounds of his voice. The wind and the clouds are his breath. Rain was made from his sweat. And from the fleas that lived in the hair covering him came all humankind.[6]

The fifth myth is from the Blackfoot American Indian people. It is called "Traveling to and from the world, Old Man the Creator."

The Old man who was the creator was traveling about making things. He had been south and was on his way north. He created the birds and the animals as he went. He made prairies, and as he traveled north, he created mountains.

He made timber and bushlands. He put red paint in the soil and he formed rivers and waterfalls.

One-day, Old Man decided to make a mother and a child. He formed them out of clay. He molded the clay in the shape of humans, and he spoke to them, "You will be people," he said. He covered up the clay shapes and went away.

The next morning, Old Man went to the place, taking the covering off the shapes. They seemed to have changed just a little. The morning after that the shapes had changed some more. The next day, they were different still.

On the fourth morning, Old Man went over and looked at the shapes that were images of people now.

"Rise up. Walk." Old Man told them, and they rose up and started walking. They, the woman and the child and their maker walked to the river.

"My name is Na'pi" he told them, which means Old Man.

The woman looked at the water and said to Old Man, "Tell me, how long will it be? Will we live here always? Will there be no end to our living?"

"Well, I haven't ever thought about it," Old Man said. "We'll have to decide. Let us take this buffalo chip and throw it into the river. If it floats, then people will die. But they will die for only four days. Four days after they die they will live once again."

He threw the buffalo chip into the water and it floated.

The woman picked up a stone. "No," she said, "let me throw this stone in the river. If it floats we will live forever, but if it sinks, we will feel sorry for one another, that we must all die"

The woman threw the stone into the river and it sank.[7]

"The First Man becomes the Devil" is the title of our sixth creation story. It is a Russian Altaic creation story in which the first man is the Devil, Erlik. Per this myth, god Ulgen saw mud floating on the waters. The mud had the shape of a human face. Ulgen gave the shape a spirit and when it lived, it was the first man. Ulgen called him Erlik.

---

[6] *Ibid.* p.21-23.
[7] *Ibid.* pp. 25-27.

In the beginning, Erlik and god Ulgen were friends. But then Erlik tried to create life of his own. He boasted about it.

"I can do as well as Ulgen. I can make man."

That made Ulgen angry. He commanded Erlik down to the depths. Now Erlik is the leader of dead spirits. He is the devil.

Ulgen next created earth. He put seven trees on it and he put a man under each of the seven trees. Ulgen made a golden mountain, and there he also placed a tree, which was the eighth one. Under this tree he put the eighth man. Ulgen named that eighth man Maidere. Then Ulgen the god went away.

Each of the seven trees grew seven branches after seven years. There was one branch for each year. But each man under each tree did not change at all. Each man stayed the same.

Ulgen returned. And asked, "Why do the men not change?"

"Well, they cannot grow and change when there are no women for them," said Maidere.

"Then come down from your golden mountain," said Ulgen. "Make women for these men."

And Ulgen walked away again.

Maidere came down. He started to create a woman. He created a body, but he could not find a way to make this first woman live.

He had to find Ulgen. And he left Dog to guard his creation. "If anyone comes, bare your teeth," Maidere told Dog. "Bark loudly and frighten him away. Don't let anyone come near the women."

"All right, just as you say," Dog said. Maidere was barely away when the evil Erlik came along. "Dog," Erlik said, "would you like a fur coat?" In those days, Dog had nothing on his skin. He was naked Dog. It was winter and cold. And he was shivering Dog."

"The coat I give you will never wear out," Erlik said. "It will feel cool in the summer and warm in the winter. And you know that any fur coat I give you will last you all your life."

"What do you want me to do in return?" asked trembling Dog.

"Just let me look at the woman Maidere has made. I only want to see her."

"Well, alright," Dog said, "As you wish."

Erlik crept close to the woman. He took out his flute and played seven tunes into her nose. Next, he played an instrument with seven strings right by her ear. The woman sat up. All at once, she was quite alive. She had a mind, and she had a spirit. But she had seven tempers and nine moods.

Dog found this out when the woman got angry for no reason and threw stones at him.

When Maidere came back home, he carried the breath of life from god, Ulged. But he was too late. The first woman didn't need the breath of life. She was already alive.

"I told you not to let anybody near the woman!" Maidere scolded Dog.

"Well, I was cold," Dog said. Erlik said he would give me a fur coat."

"In that case, I will have the fur coat grow on your back forever," Maidere said. "Let everybody throw stones at you forever and treat you badly forever, too."

"All right," sighed warm Dog.

So, it has always been. [8]

The seventh creation myth from Hamilton and Moser comes from the Krachi people of Togo in West Africa. It is entitled "Spider Ananse Finds Something".

In the beginning, was Wulbari. And god Wulbari covered mother earth about five feet above it. Wulbari was very upset. There was not enough space between the earth and Wulbari. The man who lived on the earth kept bumping his head against the god. It didn't seem to bother the man, but it bothered Wulbari.

---

[8] Ibid. pp. 29-33.

An old Woman was making food outside her hut. Her stirring pole kept knocking and poking Wulbari. The smoke from her cooking fires got into his eyes.

"I'll rise up a little bit." thought Wulbari. And so, he lifted the blue of his heavenly self just a little bit higher.

"There," he thought, "that's better."

But still being so close to woman and men, Wulbari was useful. He became a perfect towel for everybody. And the people used Him to wipe their dirty hands. There was even one woman who took a piece of clean blue to make her soup taste better.

"Ummm," she murmured. Wulbari couldn't believe it. But there it was, pieces Heaven-He, being sniffed by the dogs and eaten by the babies.

Wulbari moved up higher and higher until he was out of the way of everybody. He set up his court that included the animals, his guards, and the Spider, Ananse was their captain.

One day, Ananse asked Wulbari for a corn cob.

"Of course," Said Wulbari, "But what do you want it for?"

"Master, I will bring you a bushel of corn if you give me the corncob."

Wulbari had to laugh, and He gave Ananse the corncob. Ananse made his way down the heavenly road to the earth. He found a place to stay with a chief and asked where he could put the cob to keep it safe while he slept.

"It is the corn of god Wulbari," Ananse said. "And I must guard it."

So, the people showed Ananse a good place in the roof for safekeeping. During the night, when all were asleep, Ananse took the corn and fed it to the chickens.

The next day, Ananse made a great fuss about the missing corn. So, the chief gave him a whole bushel of corn.

That was the way it was with Ananse. He could trick all the people.[9]

The final myth we will share with you in this first look at creation stories comes from Babylonia. It is the story of Enuma Elish, perhaps the most famous of the Near eastern texts. It is entitled "Marduck, god of gods".

When there was Apsu and Tiamat, and nothing else, they created the great gods. They brought the gods Lahmu and Lahamu into being. And for ages these two grew and grew.

The god Ansher and Kishar were formed next, and they grew even taller. The god Anu was their son. He was equal to his father, Anshar.

Anshar brought the god, Ea, into being. Ea was wise, understanding and strong. H was even mightier than his grandfather. Anshar. There were to none to rival him among the gods.

The god brothers banded together in the sweet and salt waters as more of them came into being. They surged back and forth. This bothered Tiamat. Some say she was a dragon. The god sons made her moody with their noise and laughter.

Apsu could not stop the brother gods and Tiamat could not speak to them. For they were too overbearing. Apsu decided to destroy them so that he and Tiamat could have peace.

"What? Should we unmake what we have made?" Tiamat asked. Her mood was dark now. "Their ways are awful, these gods, but let us act kindly."

Apsu continued to plan evil against the gods, his sons. But the gods heard what was plotted. They became silent, all but one. He was Ea, the all-wise. Ea made a spell. He spoke the magic, and he put in the deep of the fresh water that was Aspu. His spell made Aspu fall asleep and then Ea killed him.

---

[9] *Ibid.* pp. 53-58.

Ea and Damkina, his wife, dwelled in spender in this watery place of fate; they called it the Apsu. And in the heart of the Apsu was created the majestic god, Marduk. It was Ea and Damkina's doing. They were the father and mother.

Marduk looked like a god of gods for all time. His eyes flashed and sparkled. Leader that he was, he walked like a Lord of Ages. When Ea firsts saw him, his heart was filled with rejoicing. He said Marduk was perfect and to be praised as the "most high" god.

Marduk had four eyes and four ears. When his lips moved, the fire blazed from within them. His eyes scanned everything. He was fearless and radiant. He was best and tallest, boldest and brave.

The god Anu then made the four winds. They, in turn, brought waves and foam to Tiamat's waters. Diving down, Anu filled his palm and created dirt. Waves stirred up the dirt.

Tiamat did not like being upset and so disturbed. She moved and moved, day and night. The gods could not resist. "We cannot sleep." They said. "You let Aspu be killed and did not stay by his side. Now there are four winds. You are alone. We cannot rest. You do not love us!"

"Let us make monsters then," Tiamat said.

She who could fashion all things gave birth to monster serpents. She made roaring dragons, bloodless and filled with poison and she crowned them with haloes, so they would like the gods.

Tiamat created the Viper, the Dragon, and the Sphinx, the great Lion the Mad Dog, and the Scorpion Man. She created demons. The Dragon-fly, the Centaur. There were eleven of them that she made herself. And among these creatures she made Kingu.

Kingu was the chief of the monsters and they would battle now against the fairer gods-Anshar and Ea, and Anu. They would avenge the death of Aspu. Anu went to stand against Tiamat and her terrible dark blood. But Anu could not withstand her. He had to retreat. Then Ea call his son, Marduk. And Lord Marduk was pleased. He prepared himself and stood before the fair god, Anshar.

"I will accomplish all that is within your heart," said Marduk. "I will be your avenger and slay Tiamat. But you must make me supreme. From now on, my words will fix the destinies of the gods. And whatever I create will remain unchanged."

So, the gods agreed to grant Marduk kingship of the universe. But first they spread the starry robe of the night sky in their midst.

The gods said to Marduk, "By your word, make the robe vanish."

Marduk spoke of sun and light, and the robe vanished.

"By your word," said the gods, "Let the robe appear again."

Marduk spoke in the words of night and stars, and the robe was seen again.

The gods rejoiced.

"Marduk is King!" they said.

The Marduk made ready for battle. He took up is scepter the gods had given him, his royal ring, and his thunderbolt. He took up his bow and arrow and his club. He placed lighting in front of him and made his body full of flame. Then he made a net to trap Tiamat.

The four winds helped him so that she could not get away. He brought evil winds, whirlwinds and hurricanes, to stir up the waters of Tiamat. He rode his terrifying chariot of rage. To this he tied his four-team: the Killer, the Crusher, Unyielder and Fleet.

Lord Marduk went forward wrapped in his armor. On his head was a turbaned halo. He had magic in his mouth and carried a root to protect against poison in his hand.

God Marduk called out to Tiamat to come and fight him.

Tiamat cried out in furry and cast spells. Then Lord Marduk spread his net to trap her. She screamed out a poison. Marduck countered with an evil wind. Tiamat opened her mouth to eat him, but Marduk drove the evil wind down the waterspout. He shot at arrow at her and it cut her in half.

As Tiamat was dying the monsters and demons trembled in terror. Marduk captured and smashed their weapons and then he killed each one of them. He then turned back to Tiamat and raised half of her body on high to serve as the heavens. Then he surveyed the Apsu of Ea, his father, and the deep waters. The other half of Tiamat he was to make the earth between the heavens and the water.

Marduk then made the days and the year, and the order of the planets, and the moods of the moon. He made constellations of the gods.

The one day he decided to build a creature. He told his father the creature should have a frame of bone and blood. He would call the creature "Man". This man would be created to serve the gods. [10]

---

[10] Edith Hamilton, *Mythology,* (San Diego, New York London), 1988, pp.79-99.

Section One Questions

1. What similarities did you find in these creation myths?

2. Did any of them cause you to wonder if they were possible?

3. How do you compare them to what you believe about creation?

4. Did any of them answer your questions about how things came to be?

Section Two

The First Creation Story in the Bible.

The first of the two creation stories in the book of Genesis in the Old Testament will serve as the structure or framework for the continued look at God's creation and how we humans have managed it.

To it we will add scientific theories of evolution and the life cycle to show how life has evolved over the centuries.

We begin our study of this second section by looking at the first creation story found in the *Bible*. A brief introduction will assist you in understanding why there are two creation stories and why we have chosen the first one as our guide for this book.

Among biblical scholars, the dominant view is that the Pentateuch, the first five books of the Old Testament, is a composite of several traditions that have been edited and reworked over the centuries to become what we find today. According to the *Harper Collins Bible Commentary*, "...scholars have focused on the history and development of the book. (Bible) The methods employed include source criticism,"[11] the recognition and isolation of component narrative strands. Scholars over the past two centuries have discovered that there are four major literary strands found in these first five books which they have labeled J, E, D, and P. They agree that J is the earliest of the strands, or sources, and it comes from the time of the early Jewish monarchy, around 950 B.C. It is called J because of the use of the Jewish name "Yahweh" for God in the narratives of Genesis. E relates closely to J but comes from the Northern Kingdom and is believed to be written around 750 B.C., nearly two hundred years later than J. It was called E because of the use of the name Elohim for Yahweh before the name Yahweh was revealed to Moses as God's name in Exodus 3:14-15. D is the material found primarily in the book of Deuteronomy and comes from the southern kingdom. It is identical in many ways to the law book found in the temple during Josiah's reign. It was written around 650 B.C. Finally, P is the last of the four strands. It is called P because of its priestly interests. It comes after the fall of the nation in 587 B.C. These strands were edited and pieced together in a variety of stages until the Pentateuch assumed its final form in about 400 B.C.[12]

The various inconsistencies, repetitions, and stylistic differences reflect the ways in which the story was relived, reworked, and reinterpreted in different historical periods. The P influence, for example would direct those reading the scrolls regarding worship and sacred traditions. The Deuteronic reformers tried to pull together information form the other sources and make it as clear as possible. J and E each had various stories about God and how that God interacted with their people.

Though some scholars may question the exact dates of the writings, few scholars doubt, however, that the materials from which the Pentateuch was written come from widely different periods. It is obvious to those who read carefully that the sources mentioned earlier contradict each other The material of Genesis and Exodus must have existed for centuries in some oral form, and principles of selection and arrangement that we can no longer identify shaped a good deal of the material.[13]

This theory of how the first five books of our Bible were composed helps us understand why we have two accounts of creation in the first two chapters of Genesis. According the first, (1:1-2: 4 a) man was created male and female (1:26-27), after the creation of the plants (1:11-12) and the animals (1:20-25). The second creation account is found in Genesis 2. In verses 4b-25, man was created first, then the trees, then the animals and finally the woman. These two stories are different in style, as well as fact. The first account is from the P source, the priestly source, and the second is from an edited J source that scholars call J2. So, this information tells us that the account we will be using for this study comes from the writer or writers who are concerned about carrying on the priestly traditions after the fall of Jerusalem to the Babylonians in 587 BC. These priestly accounts are found throughout the

---

[11] James L. Mays. general editor, *Harper Collins Bible Commentary*, (San Francisco), 1988, p. 83.

[12] Bernhard W. Anderson, *Understanding the Old Testament*, (Englewood Cliffs, New Jersey), 1966, pp. 16, 17.

[13] *Ibid.* p.17.

Pentateuch. It has integrity of its own but is dependent upon the J and E epic for dramatic narration. Thus, the Pentateuch finally took shape as a priestly edition of Israel's sacred history.[14]

As one might imagine, the atmosphere of worship and praise is an important theme in the P source inserts into the J and E documents. It provides us with much of the material in Leviticus and goes into detail regarding sacrifices and how the priests and people are to interact with and worship their God.

The writer or writers of P comes from within the worshiping community of Israel. They look back at the formation of a nation when God brought the people out of Egypt and from there back to the creation story. In the majesty of style and thought, the P story we will use is excelled by few others in the Bible. Its rhythms and refrains reflect what may have been years of use in worship in the Temple where it may have been recited as a part of worship until it gradually assumed its present form of liturgical prose.[15]

The author of the *Story Teller's Companion to the Bible*, puts it this way, "This first account of creation has been described as poetry, liturgy, poetic theology, a proclamatory sermon, and as many other forms of literature."[16]

There was no one present to hear God's words as God declared and things were created; no one to record that God was pleased or displeased. The account is not meant to be a literal transcript of what God said at the beginning of time. How could it be? No one was there. It was meant to communicate a theological statement. What we have in this account is not a scientific explanation of how the earth was created. It was a theological one. The ancient biblical writers were not scientists but followers of a God who had chosen their ancestors. That God brought them out of slavery and provided for them a home. Miracles occurred along the way such as the parting of the sea. Astonishing events took place such as the crumbling of the walls of Jericho. Guidance and direction were provided as God worked through Moses and Jacob to bring the Hebrew people to the Promised Land. The writers believed that these things occurred because their God was a powerful God, powerful enough to be responsible for every bit of creation.

Their understanding of the cosmos was vastly different from our own. They did not have the advantage of sailors who would circumnavigate the earth or telescopes and space stations to provide them with information about the universe beyond what their eyes could see. For them, the universe had three layers, the heavens above the earth, the earth beneath the heavens and the water under the earth.

The purpose of this creation story given us by the P source is to make it clear that everything created is dependent for its existence and meaning upon the sovereign God.

We have called this account from first Genesis a theological account of creation. We do not use that term lightly. As a college student, a beautiful African American girl, provided a rendition of these verses using only a drum, her voice and body to tell a spell-bound audience just how God created the world. It was beautiful and moving.

Walter Brueggemann, in his book, *Interpretation a Bible Commentary for Teaching and Preaching*, points to several issues that will help us as we make our way through the first creation story. The first is "more than any other part of the Bible, this material has important links to parallel literature in the ancient Near East. Not only are their parallel creation stories and flood stories, as has

---

[14] *Ibid.* p 382.
[15] Ibid. p. 384.
[16] Michael E. Williams, editor, *The Story Teller's Companion to the Bible*. (Nashville, TN.) Volume I, 1976 P. 26.

long been recognized, there are also parallels in which creation and flood are joined together in one large complex. Thus, our material relates to an old tradition even in its present shaping." [17]

As mentioned above when we talked about source criticism, Brueggemann, agrees that "More than anywhere else in Genesis, one is aware here of the problem of literary sources. It is conventional (and accepted) that these chapters are of two different traditions, commonly J and P. The J material in Genesis 2-3, 4;11:1-9, and some parts of the flood narrative and the genealogies, is usually taken to be earlier. The P source is commonly dated to the exile. It deals with the problem of despair and hopelessness. This tradition is found in Genesis 1:1-2:4a, parts of the flood narrative and elements of the genealogies." [18]

The creation story we will be looking at in greater depth is a "magnificent, formal, cadences and in balanced, seven-part structure, celebrates the creation of the heavens and the earth and all things therein." [19]

This account, as does the second account of creation in Genesis 2, seems to indicate that everything that exists was created by God in the first account, within six days. And it is assumed nothing has changed because God was pleased with what had been created.

Superimposed against this theological expression given to us by the P source, we find Darwin's theory of natural selection. Charles Darwin, one of the best-known naturalists in the world, spent five years aboard the H.M.S. Beagle, setting sail from England and traveling around the world. After stops in South America, the ship spent time in the Galápagos Islands. While in the Galápagos, Darwin observed that the mockingbird species (Image 2a), while similar to those in South America, differed slightly from island to island. The same held true for the tortoises (Image 2b) on the islands, with shell sizes and shapes differing from island to island. Along with the mockingbirds and tortoises, Charles Darwin based his theory of natural selection on the finches found in the Galápagos Islands. Charles Darwin's finches are considered one of the most famous, if not the best-known group of birds in the world. In total, there are fourteen species of the subfamily Geospizinae, thirteen of which are found in the Galápagos Islands, and one found on Cocos Island. [20]

---

[17] Walter Brueggemann, *Interpretation A Bible Commentary for Teaching and Preaching (Nashville)*. 1976 p.14.
[18] Ibid. p.14
[19] Edward P. Blair, *Abington Bible Handbook*. (Nashville), 1975. p.91.
[20] Peter R. Grant, *Ecology and Evolution of Darwin's Finches* (Princeton, New Jersey: Princeton University Press, 1986).

(Image 2.a. Galápagos Mockingbirds)

(Image 2.b. Giant Tortoise in the Galápagos Islands)

Darwin spent two years in the Galápagos Islands, and while he only observed nine of the thirteen Galápagos finch species, it was enough time for him to draw his conclusions on natural selection and evolution.[21] Through his observations, Darwin concluded that the finches in the Galápagos Islands were similar enough to have come from a common ancestor. This is reflected in one of his manuscripts as follows: "[t]he most curious fact is the perfect gradation in the size of the beaks of the different species of *Geospiz* – Seeing this gradation and diversity of structure in one small, intimately related group of birds, one might really fancy that, from an original paucity of birds in this archipelago, one species had been taken and modified for different ends.[22]"

The beak size, shape, and use afford the finches the opportunity to exploit unique niches allowing them to coexist in the islands. Specific finches use their beaks of various sizes for specific activities and are often found in unique habitats. Though the fourteen-finch species are unique species, their habitats and beak size may overlap. For example, from Figure 2.a which lists ten of the fourteen finch species, it is clear the different beak sizes are associated with different species. The beak sizes

---

[21] Peter R. Grant, *Ecology and Evolution of Darwin's Finches* (Princeton, New Jersey: Princeton University Press, 1986).
[22] Charles R. Darwin, *Journal of Researches into the Geology and Natural History of the Various Countries Visited During the Voyage of H.M.S. 'Beagle', Uunder the Command of Captain FitzRoy, R. N. from 1832 to 1836* (London: Henry Colborn, 1842).

and shapes are also associated with certain food groups such as vegetarian, insect eaters, cactus feeding, and seed eaters.

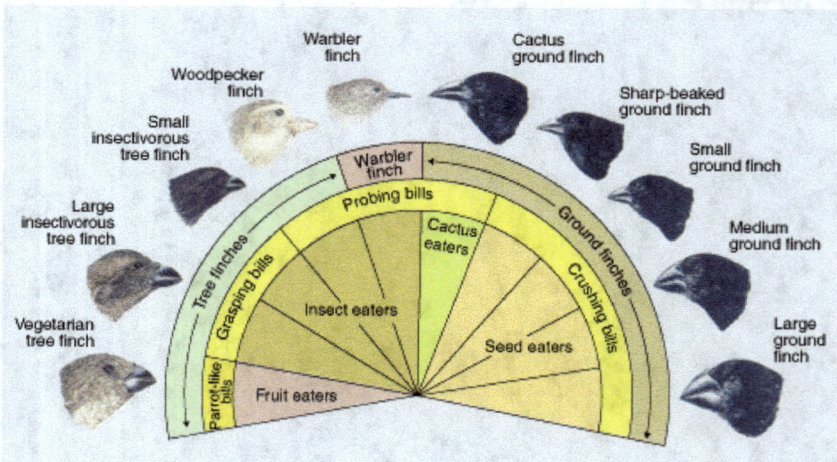

(Figure 2.a. Beak sizes of Darwin Finches[23])

When comparing the varying ground finches, it is easy to see where they get their name. The large ground finch, which is a seed eater, uses its large beak to crush seeds. The medium and small ground finches are aptly named based on their sizes. The sharp-billed ground finch feeds on seeds and uses its beak to peck at the feathers of blue-footed boobies in an attempt to draw and drink blood. The woodpecker finch, an insect eater, uses its long slender beak to pick up twigs and probe trees for insects.

The Darwin finches are also specific to certain habitats. All the finches spend time feeding on the ground and in the vegetation, but cactus finches spend more time in cactuses than other finches. Ground finches are found more often on the ground than the others, and tree finches are found more often in trees than other finches.[24]

---

[23] PBS website, accessed January 27, 2007, http://www.pbs.org/wgbh/evolution/library/01/6/l_016_02.html.
[24] Peter R. Grant, *Ecology and Evolution of Darwin's Finches* (Princeton, New Jersey: Princeton University Press, 1986).

(Image 2.c. Medium ground finch, *Geospiza fortis*, in the Galápagos Islands)

Following his visit to the Galápagos Islands, Darwin formulated his theory of natural selection, eventually publishing the theory as part of *On the Origin of Species* in 1859, twenty-three years after the voyage on the H.M.S Beagle. Charles Darwin defines natural selection as "the preservation of favorable variations and the rejection of injurious variations.[25]" Natural selection is a slow-moving process in which the organisms best adapted to their environment survive and pass on those genetic characteristics that help them survive to the next generation, while the less adapted organisms are naturally removed from the population (see Table 2.a).

---

[25] Charles R. Darwin, *The Origin of Species* (New York: Barnes and Noble Books, 2004).

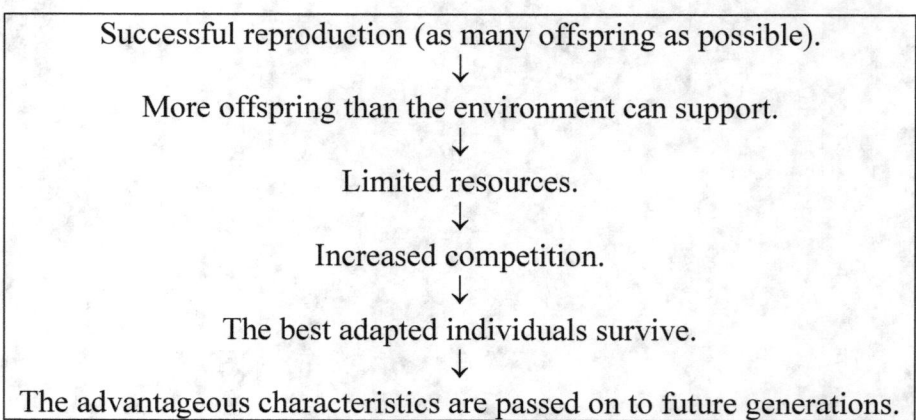

(Table 2.a. Natural Selection)

Along with mutation (change in the DNA sequence), genetic drift (the process of random changes in the proportions of inherited traits), and gene flow (the process of incorporation of genes from one population to another), natural selection is an important aspect of evolution. Defined as a theory that plants and animals have their origin in other preexisting types of plants and animals, evolution suggests that all species arose from a common ancestor. Using the analogy of a tree, Darwin suggests that "the green and budding twigs may represent existing species; and those produced during each former year may represent the long succession of extinct species (Darwin, 2004)." He relays that while there were once many twigs, only a few large branches now remain which support the many other twigs and branches that have developed over time. In the end, "the great Tree of Life ... fills with its dead and broken branches the crust of the earth, and covers the surface with its ever branching and beautiful ramifications.[26]"

A well-studied example of natural selection is the peppered moth population in the United Kingdom. Peppered moths are of both light and dark colored morphs (Image 2d). During the industrial revolution, soot covered the trees in northern Europe giving the dark colored moths the ability to blend in with the environment and making the light-colored moths more susceptible to predation. Eventually, nearly all of the moths were of the dark colored morph, since the light-colored morph either did not survive to reproduction, or produced offspring which were of the dark colored morph. With the improvement on air quality, the soot on the trees dissipated giving the few light-colored moths an advantage, leading to more frequent light-colored offspring and increased predation on the dark colored moths. Currently, the majority of the peppered moths in northern Europe are of the light-colored morph.[27]

---

[26] Charles R. Darwin, *The Origin of Species* (New York: Barnes and Noble Books, 2004).
[27] Robert L. Smith and Thomas M. Smith, *Ecology and Field Biology (Sixth Edition)*, Menlo Park, California: Benjamin Cummings, 2001), pp.364-365.

(Image 2.d. Peppered Moths[28])

As natural selection is a slow-moving process, taking several generations for visible signs of change, it is difficult to observe the process in real life situations. The isolated Galápagos Islands were an ideal location for Charles Darwin to develop his theories, as various habitats on the islands held similar, but unique species. Though natural selection and evolution have been scientifically proven, and can be replicated in laboratory settings, the theories remain controversial. In many cases evolution has been prohibited from being taught in school since it does not conform to the literal meaning of religious texts. For additional reading related to Supreme Court cases involving evolution, please see the following website: http://ncse.com/taking-action/ten-major-court-cases-evolution-creationism.[29]

These scientific understandings of creation and evolution provide us with additional insight in to how fragile and complex our earth is and how important it is that we who are the superior species on this planet see clearly the need to control the destruction of it through policies and practices that are detrimental to any one part of it.

---

[28] Truth in Science website, accessed May 25, 2011, http://www.truthinscience.org.uk/site/content/view/127/65/
[29] National Center for Science Education website, accessed October 27, 2017, http://ncse.com/taking-action/ten-major-court-cases-evolution-creationism

Section Two Questions

1. Do you understand the theory of authorship of the first five books of the Bible?
2. Does it make sense that the Priestly source would be the source of the second creation story?
3. Can you imagine that animals and plants have adapted and evolved over the centuries?
4. Is it conceivable to credit God with creation as well as the ability for plants and animals to evolve?

Section Three

Biblical Examination

Let us begin this section of the study by looking more closely at the Biblical verses that make up this first creation story. There were seven days in this account, but only on six was there any creation. On the seventh, it is said that God rested.

We will begin each chapter in this section by reviewing the scripture and follow that by adding any scientific information that we find relevant and then conclude with suggestions on how we may be able to make positive changes to our environment.

Adam Hamilton says in his book, *Making Sense of the Bible*, "The consensus view is that the universe as we know it had its beginning 13.77 billion years ago, the earth is approximately 4.5 billion years old. Dinosaurs as we typically think of them were around from 230 million to 66 million years ago, the first anatomically modern humans are seen in the fossil record about 200,000 years ago, art begins to show up on cave walls about 40,000 years ago, the last Ice Age ended about 12,000 years ago. This brings us to a period called the Neolithlic, sometime called the "Agriculture Revolution" when human beings began to move from being hunter-gathers, to raising livestock and farming. Cities began to emerge around 9,000 BC in the region of Mesopotamia and the rest of the Fertile Crescent.[30]

James Mays shares in the *Harper Bible Commentary* that "The six days of P's creation work are described in a solemn, repetitive language and are arranged in two set of three, in which the first and fourth days correspond, as do the second, and the fifth and the third and the sixth. The first three days describe the divine work of separation (light from darkness, water above and water below, sea from land) that prepares a habitable world; the last three days describe the inhabitants. The first day describes the creation of light, and the fourth day, the source of the light (sun, moon, stars); in the second day, the creation of the dome of the sky ("firmament" in the older translations) provides an air space and water for the creatures of the fifth day, fish and birds; for the third, day land and plants are provided for the land animals and human beings of the sixth day." [31]

---

[30] Adam Hamilton, *Making Sense of the Bible*, (New York, New York), 2014 p.12.

[31] James L. Mays, general editor, *Harper Collins Bible Commentary*. (San Francisco), 1988, p.86.

Now that we have given some Biblical and scientific background, let's begin to look at each of the six days of creation. We will examine what the verses tell us, try to explain what was intended and then, try to explain what science tells us that relates to the day in question.

Day One: Light

Gen.1:1-5

[1] In the beginning God created the heavens and the earth.
[2] The earth was without form and void, and darkness was upon the face of the deep; and the Spirit of God was moving over the face of the waters.
[3] And God said, "Let there be light"; and there was light.
[4] And God saw that the light was good; and God separated the light from the darkness.
[5] God called the light Day, and the darkness he called Night. And there was evening and there was morning, one day.

As you have seen in the first section of this book, there are many varieties of creation stories. Nearly every ancient culture had one. We have shared a few with you in the first section of this book. They served to answer the question of where we came from and how the stars got in the sky. The P writer was influenced in the writing of his creation story by an earlier Babylonian story. This accounts in part for an already existing darkness and chaos. But to counter the notion that the chaos did not come from God, the writer begins his account with "In the beginning".[32] In the beginning, before there was chaos, God created the earth and everything else the people could see in the sky.

The *Harper Bible Commentary* says, "As most modern translations recognize, the P creation account (1:1-2:4a) begins with a temporal clause ("When in the beginning, God created"): such a translation puts Gen. 1:1 in agreement with the opening of the J account (2:4a) and with other ancient Near Eastern creation myths."[33]

The *Harper Collins Bible Commentary* offers several bits of information that will help us better understand this first chapter of Genesis. "The Hebrew word for 'create' is used exclusively of divine activity; heavens and earth are a merism (an expression of totality by the use of polar expressions) meaning "everything…"[34]

In verse two the reference to the "Spirit of God" is a latter addition to the original account written by P. It was added to try to explain that God or God's Spirit, was acting in the midst of nothing. It stands out because all of the rest of creation occurs with the uttered word of God. In this verse, only, God's Spirit is a work.[35] Many people have tried to read into this phrase that this implies the existence of a trinity, but that argument holds no ground. The phrase did not say "They". It refers to the spirit of God which is not something we find in scripture in this same sense until much later in the book of Samuel.

Mays offers his opinion regarding verse two by saying, "The description of the precreation state in verse two probably is meant to suggest a storm-tossed sea: darkness, a great wind and watery abyss."[36]

---

[32] *Interpreters Bible Commentary*, Volume I, (Nashville, 1952) p. 467.
[33] James L. Mays, editor. *Harpers Bible Commentary*, (San Francesco, 1988).p. 87.
[34] Harper Collins Bible Commentary. San Francesco. James L. Mays, editor. 1988 p. 86.
[35] *Interpreters' Bible Commentary*, Volume I, p. 466.
[36] *Harper Collins Bible Commentary*, (San Francisco. James L. Mays. Editor. 1988). p.86.

Verse three sets the tone for the rest of this creation story. God speaks, and so it is. The power to create comes in the spoken word. "Let there be light." And so, it was. Scholars have also found that light was also the first thing to be created in the creation accounts of the Indian, Greek and Phoenician cosmogonies.[37]

One would think that the first day of creation would be the most important one. It set the rest in motion. It served as the foundation for the rest. The earth that had just been formed was far from complete. God still had much to do with the universe. At the top of the agenda was light. Light is essential to life. So, before there were plants or animals, there had to be light. But this verse does not describe the sun and the stars as we might expect. They are found in verse 16 on what is the fourth day of this account. But for some reason, P has light created here on this first day, before the sun. We are not sure there is an explanation for this and it certainly is baffling. Perhaps it goes back to the advantage we have of looking back through a history filled with scientific revelation and knowledge that the P writer simply did not have. He must have known or believed somehow that light was essential to life, perhaps he had heard one of the other accounts that other cultures had of creation that included light being the beginning of all things. Whatever the reason, P has the creation of day and night occurring appropriately in the beginning of creation.

As we examine this creation of light, it occurs to us that centuries after this was written, the author of the gospel of John in the New Testament also uses light as he introduced God's activity in the world. John says in the first few verses of the book, "In the beginning was the Word, and the Word was with God, and the Word was God. He was in the beginning with God. All things came into being through him, and without him not one thing came into being. What has come into being in him was life and the life was the light of all people. The light shines in the darkness, and the darkness did not overcome it."

Perhaps the P source is trying to make a theological statement, much like John did. God was the creator and part of what was created was goodness and life in the form of light that cannot be overcome.

We often associate light with goodness and dark with evil. The goodness God provides in the light may be the first gift given in creation.

In verse four, God sees the result of the spoken word, and the verse says, "God saw that the light was good." Perhaps we could imagine that God was pleased that what had been created was just what was intended. There was light. The scripture does not indicate why there had to be night, but that is the next thing to be created. I am sure the P source did not know the earth orbited the sun, so he did not understand why there was day, light and night, darkness. And as a result, the next step was to separate light from darkness. There was nothing spoken to make this happen. God simply did it. And the light was called day and the darkness was called night.

We would like to provide some information regarding our solar system and what we know about day and night. Our solar system contains the Earth, the sun, seven other planets, and other celestial bodies, and is part of the Milky Way galaxy. It is believed that there are 100 thousand million stars in our galaxy, and that there are 100 thousand million galaxies in the universe. Therefore, the current estimate is that there are ten sextillion (10^{22}) to one thousand sextillion (10^{24}) stars in the universe.[38]

---

[37] Ibid p. 469.
[38] European Space Agency website, accessed February 16, 2016,
http://www.esa.int/Our_Activities/Space_Science/Herschel/How_many_stars_are_there_in_the_Universe

Where did all these stars, planets and associated bodies come from? One theory is the Big Bang Theory, which is defined as the singularity at the beginning of the universe. A singularity is a point in space-time that is infinite. Simply, the big bang is the beginning of time.

Alexander Friedmann, a Russian cosmologist, suggested in the 1920s that at some point in the past, between ten and twenty thousand million years ago, the distance between neighboring galaxies must have been zero. At the point of the big bang, the universe had zero size and was infinitely hot. As the universe expanded, temperature decreased by ten thousand million degrees colder within one second after big bang. In the beginning, the universe contained mostly photons, electrons, and neutrinos, their antiparticles, protons, and neutrons. One hundred seconds after big bang, temperature would have cooled enough (to one thousand million degrees) to allow the formation of helium nuclei (made of protons and neutrons), hydrogen, and some other elements. Production of helium and other elements would have stopped within a few hours after the big bang.

For about a million years, little happened as the universe continued to expand. Once the temperature dropped below a few thousand degrees, electrons and nuclei would not have enough energy to overcome their electromagnetic attraction, thus combining to form atoms. As the universe continued to expand and cool, the denser regions slowed down due to gravitational attraction. This led to a reversal of expansion in the denser regions, which allowed them to collapse and start rotating. As the collapsing regions became smaller, they spun faster and faster leading to today's galaxies.

As hydrogen and helium gas broke up into smaller clouds, they collapsed under their own gravity. The atoms in the clouds collided with each other producing enough heat to start nuclear fusion reactions which converted hydrogen into more helium. Pressure was increased from the additional helium, stopping the collapse of these clouds, allowing them to remain stable forming stars such as our sun, which was formed approximately five thousand million years ago.

Most of the gas that was in the cloud that formed our sun either went into forming the sun or was blown away by various reactions. However, a small number of heavier elements collected together to form our planets and other bodies that orbit the sun. Earth was initially very hot with no atmosphere, but it cooled over time acquiring atmosphere from gaseous emissions from rocks. However, there was no oxygen in the atmosphere. The first forms of life on earth consumed minerals such as hydrogen sulfide and released oxygen. Over time, enough oxygen built up in the atmosphere, allowing for the development of higher life forms including humans.[39]

This theory of creation explains how the earth came to be. Additional information comes to us from the book *How the Earth Works*, by Clive Gifford. According to Mr. Gifford, Earth's story began about 4.6 billion years ago, shortly after the sun started forming. Dust, ice and rock orbiting the early sun clumped together in larger and larger lumps that collided with one another, increasing in heat and size.

One of these, earth, grew until it was large enough for its own force of gravity to attract further dust and gas. As the collisions continued, generating huge amounts of heat, the early Earth's surface was repeatedly melted and reshaped. Gradually, materials rich in iron were drawn toward Earth's center, forming a large dense core of iron with lighter rock forming a thick layer, known as the mantle, around the iron. Over many millions of years, the planet cooled, the atmosphere developed, and rain fell, eventually forming lakes, seas and oceans.

Single-celled algae appeared in the sunlit areas of the oceans as early as two billion years ago. The sunlight helped create plants that produced oxygen. In time, animals and humans came into being.[40]

---

[39] Stephen A. Hawking, *A Brief History of Time* (New York: Bantam, 1988).
[40] Clive Gifford, *How the World Works* (London: Kingfisher, 2013).

About the same time that Friedmann was sharing his theory, a Belgian priest named Georges Lemaître theorized that the universe began from a single primordial atom. The idea subsequently received major boosts by Edwin Hubble's observations that galaxies are speeding away from us in all directions, and from the discovery of cosmic microwave radiation by Arno Penzias and Robert Wilson.

The glow of cosmic microwave background radiation, which is found throughout the universe, is thought to be a tangible remnant of leftover light from the big bang. The radiation is akin to that used to transmit TV signals via antennas. But it is the oldest radiation known and may hold many secrets about the universe's earliest moments.

Warmth, daylight and energy all come to mind when one thinks of the sun (Image 3.1.a). The earth is unique in that it is positioned an appropriate distance from the sun to not be too hot or too cold to sustain life. The atmosphere and ozone help protect Earth from harmful radiation so that the planet is habitable. Distance from the sun, our atmosphere, and the sun's radiation all make the Earth uniquely habitable, and one wonders if there are similar planets in the universe? A program developed in 2009 uses the NASA's Kepler space telescope to try to answer that question by attempting to locate Earth-like planets. The program, detects planets through changes in brightness when planets cross in front of a star. More than 2,700 potential planets were located in the first four years of the telescope's use, with an estimate that 90% of those will be confirmed as being planets. Once a planet is confirmed, scientists determine if these planets are located within a habitable zone. The telescope doesn't search for signs of life on a planet, but future technologies may one day allow for such exploration.[41]

Weather there was a big bang or a creation that came about as the result of God's spoken word, the fact remains that here we are. We find ourselves on this planet and have been given the responsibility to protect it.

---

[41] Science.com website, accessed August 28, 2013, http://www.space.com/20720-earth-like-alien-planets-discovery.html.

(Image 3.1.a. Sunset in the Galapagos)

    The sun's radiation provides warmth to Earth, but greenhouse gases such as carbon dioxide, methane, nitrous oxide, halocarbons, and water vapor play a role in the Earth's climate as they trap heat that reaches the Earth's surface in the form of radiation. Without these greenhouse gases, the Earth's surface temperature would be approximately 60°F colder. The amount of greenhouse gases in the atmosphere has increased over recent years leading to increased heat-trapping, which in turn altered the Earth's climate. As illustrated in Figure 3.1.a, global surface temperature has consistently increased while the amount of radiation reaching Earth's atmosphere has remained consistent.

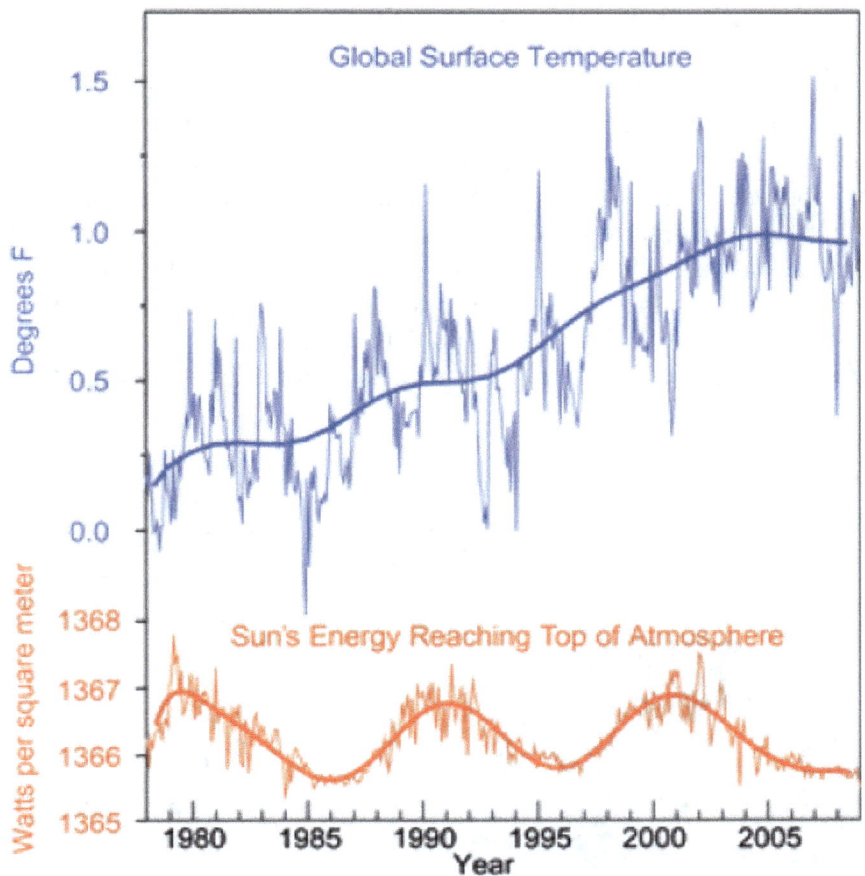

(Figure 3.1.a. Global surface temperatures versus radiation.[42])

World-wide temperatures have been increasing in recent years, with the decade of 2001 through 2010 being the warmest decade on record. In fact, the global temperature from 2001-2010 was 0.21°C warmer than 1991–2000, which in turn was +0.14°C warmer than 1981-1990. Every year between 2001 and 2010 was at the time one of the ten warmest years on record, with the exception of 2008.

Above-average temperatures were observed throughout the world, with more national temperature records broken during the decade than in any previous decade. The greatest temperature increases were observed at the higher latitudes in the northern hemisphere. For example, Greenland recorded the largest increase over the decade with a +3.08°F average and the largest increase of all countries in 2010 at +5.76°F above average.

January 2016 was the warmest January on record while also being the highest above average month on record. January was over 2°F above normal, which is higher than the previous record of December 2015 at slightly below 2°F above average. January of 2015 marked the fourth consecutive month where the global temperature was more than 1.8°F above normal. October 2015 through

---

[42] Thomas R. Karl, Jerry M. Melillo, and Thomas C. Peterson (eds.), *Global Climate Change Impacts in the United States* (New York: Cambridge University Press, 2009).

January 2016 are the only four months where the globe has topped that mark since record keeping began, 135 years ago.[43]

Several reports suggest that the likely cause for temperature increases over the last 50 years is human induced warming. The primary cause of warming is increased greenhouse gases where carbon dioxide emissions represent 77% of worldwide greenhouse gases, methane emissions represent 14%, nitrous oxide emissions represent 8%, and other gases represent 1%. Carbon dioxide present in the atmosphere is the most prevalent greenhouse gas, with emissions from practices such as burning of fossil fuels such as oil, coal, and natural gas. As of the end of 2011, the United States ranked second in global carbon dioxide emissions with the top five countries being China (29% of global carbon dioxide emissions), the United States (16%), the European Union (11%), India (6%) and the Russian Federation (5%). Figure 3.1.b is a breakdown of carbon dioxide emission sources in the United States of America.[44,45]

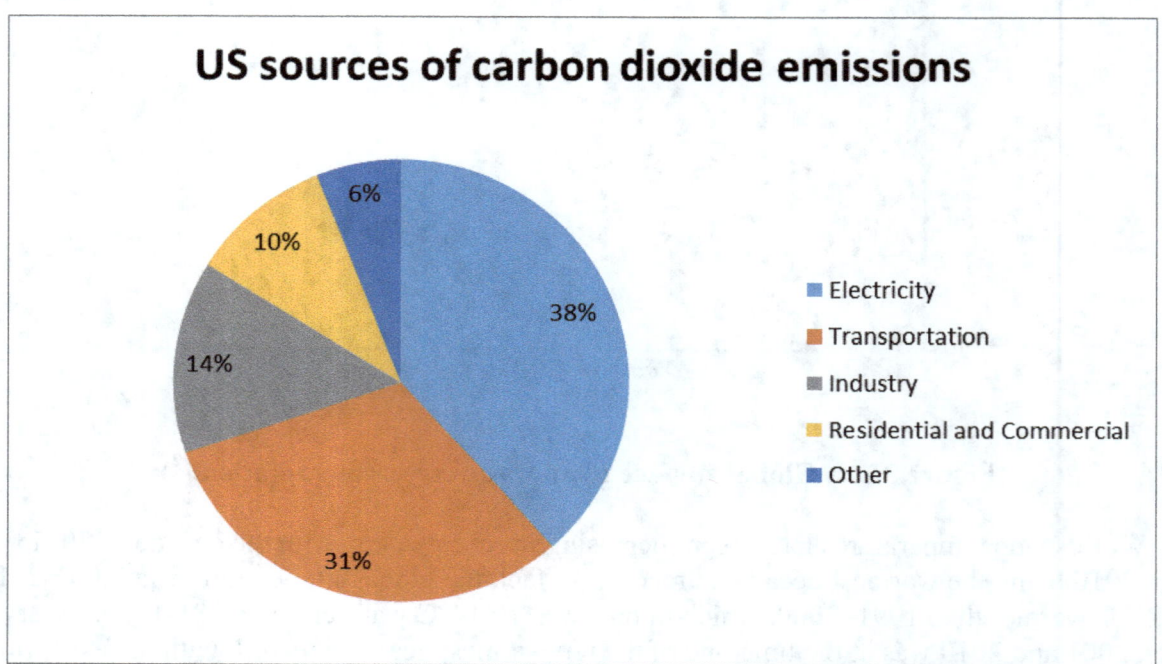

(Figure 3.1.b. United States sources of carbon dioxide emissions)

Atmospheric carbon dioxide has increased by 35% since the onset of the industrial revolution in the mid-1700s, with current levels at an all-time high (see Figure 3.1.c). For the past 250 years, humans have relied on fossil fuels for many activities including transportation, electricity generation, industrial uses, and household uses such as cooking and heating. Burning of fossil fuels has led to 75% of human-induced increases in carbon dioxide emissions, while 20% has come from deforestation and agriculture.

---

[43] msn website, accessed February 16, 2016, http://www.msn.com/en-us/weather/topstories/january-smashed-another-global-temperature-record/ar-BBpzSlc?li=BBnb7Kz.
[44] Jos G.J. Olivier, Greet Janssens-Maenhout, and Jeroen A.H.W. Peters, *Trends in global CO2 emissions; 2012 Report* (The Hague/Bilthoven: PBL Publishers, 2012), 40 pp.
[45] United States Environmental Protection Agency, *Inventory of U.S. Greenhouse Gas Emissions and Sinks: 1990-2011*, https://www.epa.gov/sites/production/files/2015-12/documents/us-ghg-inventory-2013-main-text.pdf.

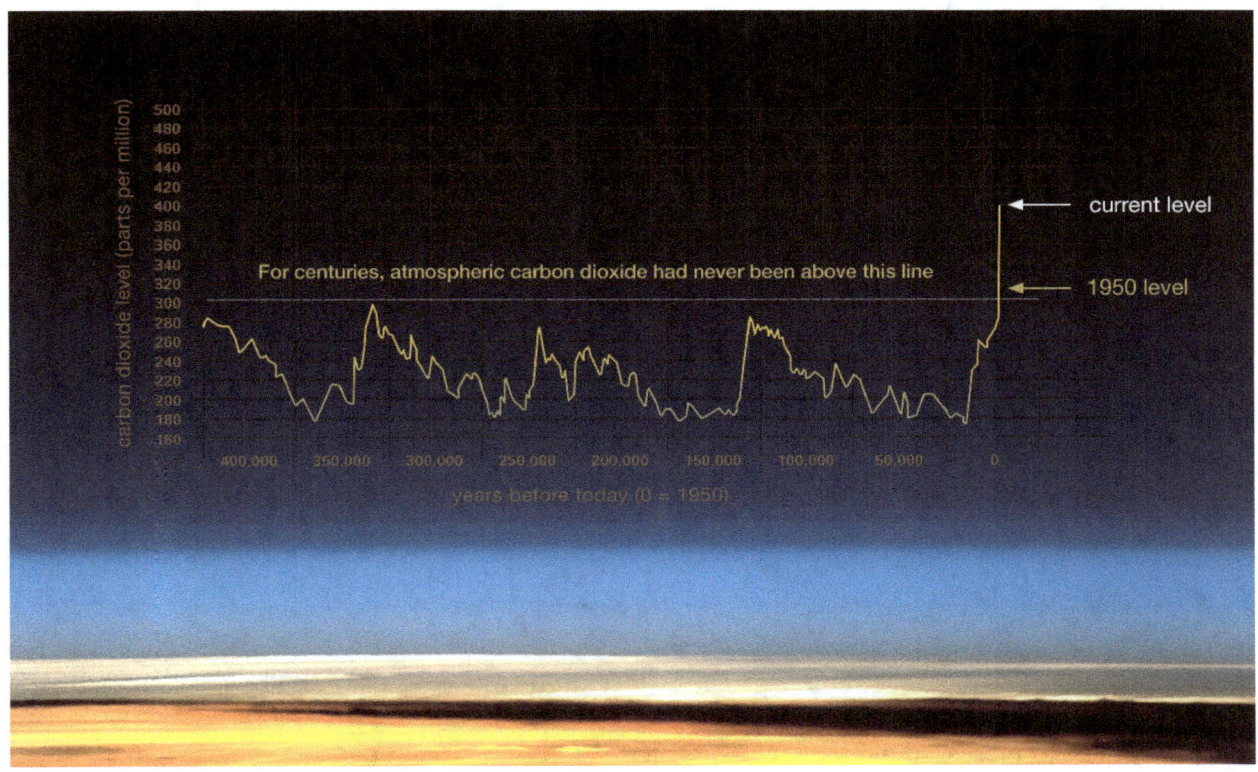

(Figure 3.1.c. Historic atmospheric carbon levels[46])

Deforestation plays a role in global warming as plants take up carbon dioxide, thereby reducing atmospheric amounts. Forest soils are often moist, but they dry out when baked by the sun after deforestation. Additional information on deforestation can be found in Day Three.

Additional human activities tied to warming include emissions of greenhouse gases including methane and nitrous oxide. Seventy percent of methane emissions can be tied to human activities including mining, transportation, sewage, landfills, and raising livestock that produce methane such as cattle. Fertilizer use and burning of fossil fuels has led to increases in nitrous oxide. Figure 3.1.d illustrates changes in atmospheric carbon dioxide, methane, and nitrous oxide over the past two millennia, with exponential increases observed beginning in the 1880s.

---

[46] NASA website, accessed March 8, 2017, https://climate.nasa.gov/climate_resources/24/.

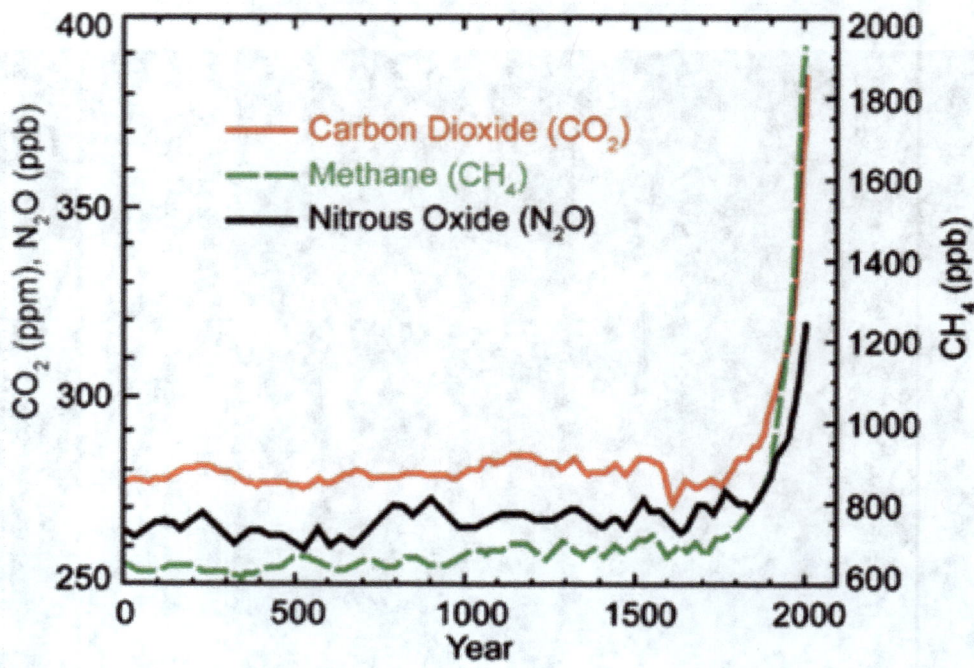

(Figure 3.1.d. Greenhouse gas concentrations over the past 2000 years.[47])

In addition to increased average temperatures, we are now seeing increases in extreme temperatures. Gutowski et al (2008)[48] suggest that "future changes in extreme temperatures will generally follow changes in average temperature." For example, high temperatures and heat waves have started to increase. Nearly half of the countries in the world reported record nationwide temperatures between 2001 and 2012, which was 25% higher than 1991 through 2000. In addition, between 2001 and 2010, more than 370,000 people died as a result of extreme weather and climate conditions, including heat waves, cold spells, drought, storms and floods, which was 20% higher than the previous decade. The main driver for weather related deaths during the decade was heat waves in Europe in 2003 and Russia in 2010 that led to 136,000 deaths. In 2012, the hottest summer on record and the most extreme drought in 50 years was observed in the United States.

Related to high temperatures, cold days and cold nights are likely to decrease, as are nights with frost. Record cold temperatures were reported in approximately one third of the countries worldwide between 1961 and 1970, while only 11% of countries reported record cold temperatures between 2001 and 2010.

Researchers indicate that these trends are likely to increase over this century, with an estimated average temperature increase between 2 and 11.5°F. If emission of greenhouse gases is cut

---

[47] Thomas R. Karl, Jerry M. Melillo, and Thomas C. Peterson (eds.), *Global Climate Change Impacts in the United States* (New York: Cambridge University Press, 2009).
[48] William J. Gutowski, Jr., Gabriele C. Hegerl, Greg J. Holland, Thomas R. Knutson, Linda O. Mearns, Ronald J. Stouffer, Peter J. Webster, Michael F. Wehner, and Francis W. Zwiers, "Causes of Observed Changes in Extremes and Projections of Future Changes" in *Weather and Climate Extremes in a Changing Climate. Regions of Focus: North America, Hawaii, Caribbean, and U.S. Pacific Islands*, ed. Thomas R. Karl, Gerald A. Meehl, Christopher D. Miller, Susan J. Hassol, Anne M. Waple, and William L. Murray (A Report by the U.S. Climate Change Science Program and the Subcommittee on Global Change Research, Washington, DC, 2008), p. 81-116.

substantially, the increase would be on the low end of the range. However, if emissions remain stagnant or even continue to rise, the temperature increase would closer to the estimate of 11.5°F.[49]

Reducing greenhouse gas emissions, specifically carbon dioxide emissions, would slow the rate of global warming in coming years, and may help to bring the global temperatures back to historic levels. According to Woodcock et al (2009)[50], 75% of transportation related emissions come from vehicular traffic. Emissions from transportation related activities are expected to rise dramatically over the next twenty years, with a predicted increase of 80% between 2007 and 2030. Seventy-five percent of transportation related emissions come from vehicular traffic. There are many ways that you can help reduce carbon dioxide emissions from transportation related activities.

Questions for day one:

1. Compare the Bing Bang Theory with what we find on day one and following of the creation account in Genesis, chapter one.
2. How are they similar and different?
3. What can we do to reduce greenhouse gas emissions?
4. What other measures can we take to live more efficiently with less carbon combustion?

---

[49] World Meteorological Organization, *The Global Climate 2001-2010: A Decade of Climate Extremes - Summary Report* (Geneva: World Meteorological Organization, 2013), http://library.wmo.int/pmb_ged/wmo_1119_en.pdf.
[50] James Woodcock, Phil Edwards, Cathryn Tonne, Ben G Armstrong, Olu Ashiru, David Banister, Sean Beevers, Zaid Chalabi, Zohir Chowdhury, Aaron Cohen, Oscar H. Franco, Andy Haines, Robin Hickman, Graeme Lindsay, Ishaan Mittal, Dinesh Mohan, Geetam Tiwari, Alistair Woodward, Ian Roberts, "Public Health Benefits of Strategies to Reduce Greenhouse-gas Emissions: Urban Land Transport," *Lancet* 374 (2009): 1930–1943.

Day Two: The Firmament

Genesis 1: 6-9

[6] And God said, "Let there be a firmament in the midst of the waters, and let it separate the waters from the waters."

[7] And God made the firmament and separated the waters which were under the firmament from the waters which were above the firmament. And it was so.

[8] And God called the firmament Heaven. And there was evening and there was morning, a second day.

The term "Firmament" denotes the expanse stretched across the sky to separate the upper and lower waters. The Hebrew word means a "strip of beaten metal" and goes back to the understanding that the sky is a mirror-like surface. [51] The Story Teller's Companion to the Bible calls the firmament a "vault".[52]

Thomas Shepard agrees and says that the word firmament is associated with firmness and solidity. He also says that it is often seen as a "faulted solid body". The term could also mean "expanse".[53]

In this second action of creation, we once again find similarities to the Babylonian myth of creation. According to this myth, the sun-god Marduk split the slain chaos monster in two and used one half of the carcass to make the firmament and the second half to make the earth.[54]

Once again, we find God creating through the spoken word. Above the firmament was the heavenly sea we find references to in Psalm 29 and Revelations 4. The author must surely have experienced rain and thunder and believed that the water that came with the rain came from the heavens and it was similar to the water that he knew on earth. It surely was not much of a stretch to conclude that they had been one in the same at some point and that the all-powerful creating God had separated them.

There are references to God opening windows or doors to allow the water to fall to the earth in the form of rain.

The second day creates a separation between two water forms. The water on earth was separated from what the writer believed was water above it.

Thomas Shepard writes a chapter in the book "The Genesis Creation Account and Its Reverberations in the Old Testament." "So, on the first day, light was created, and it was separated, distinguished from darkness." Shepard asserts that the widespread notion that the biblical cosmology reflected a pagan picture of the three-storied universe. He defends this by noting that the word "Deep" in Genesis 1 and 2 figures prominently in the argument of those scholars supporting the verses that the Genesis cosmology is three-storied. There is heaven above and earth below. Under the earth is the "deep", interpreted as the "Primeval ocean."

---

[51] *The International Dictionary of the Bible*, (Nashville, 1962) Volume 2, Nashville. p. 270.
[52] Michael E. Williams, editor, *The Story Teller's Companion to the Bible*, (Nashville, 1976, p. 25.
[53] *The Genesis Creation Account and Its Reverberations in the Old Testament*, (Berrien Springs, Michigan, 2015) p. 19.
[54] *The Interpreter's Bible*, (Nashville, 1952) Volume I, p. 472.

Mr. Shephard asserts that it has been claimed that the term "deep" is directly derived from the name Tiamat, mentioned earlier, who was the sea mythical Babylonian monster and goddess of the primeval ocean world in the national epic "Emuma Elish". "Deep" is said to contain an "Echo of the old cosmogenic myth", in which the creator, god Marduck engages Tiamat in combat and slays her.[55]

Despite this apparent relationship to the Babylonian creation myth, some scholars, including Mr. Shepard, believe that the thirty-five images of "Deep" and its derivative forms found in the Old Testament show us that it is a "poetic term for a large body of water", which is completely nonmythical.[56]

As mentioned earlier, the P source wrote this account while the Jews were in captivity in Babylonia where he or they could have heard the story and incorporated pieces of it into their own account that we now find in Genesis, chapter one.

We can break our discussion of water into two major categories: oceans and precipitation. Within the ocean category we have topics such as hurricanes, melting sea ice, and sea level rise. Within the precipitation category we have topics including rainfall, flooding, and drought.

Oceans cover over 70% of the Earth's surface, contain over 97% of the water on Earth, and influence climate and weather patterns. The five oceans (Atlantic, Artic, Indian, Pacific, and Southern) hold nearly a quarter-million known species, while scientists believe that there may be millions more that have yet to be discovered. The oceans are vast and deep, with an average depth of 4000m, but most life is located near the surface as light only penetrates to ~100m, in regions where nutrients are rich due to upwellings or mixing of currents and/or warm and cold waters, and near the coastlines.[57]

The fact that coastlines are one of the concentrated areas of ocean life is concerning when considering that coastal waters are starting to experience the effects of climate change. Rising temperatures have led to thawing sea ice, retreating glaciers, rising sea level, and lengthened growing seasons. Other climate change related affects include increased heavy downpours which leads to soil and chemical runoff into oceans and bays, and increased hurricane frequency and intensity.[58] According to Biello (2008), more than a third of the ocean waters have been seriously impacted by humans, with at least 96% of all ocean waters being impacted to some extent.[59]

---

[55] Thomas Shepherd, author of one chapter in *The Genesis Creation Account and Its Reverberations in the Old Testament*, (Berrien Springs, Michigan), pp. 15-16.
[56] Ibid. p.18.
[57] Robert L. Smith and Thomas M. Smith, *Ecology and Field Biology (Sixth Edition)*, Menlo Park, California: Benjamin Cummings, 2001), 364-365.
[58] Thomas R. Karl, Jerry M. Melillo, and Thomas C. Peterson (eds.), *Global Climate Change Impacts in the United States* (New York: Cambridge University Press, 2009).
[59] David Biello, "Ocean Impact Map Reveals Human Reach Global As vast as the oceans are, almost no waters remain untouched by human activities," *Scientific American* (February 15, 2008) http://www.scientificamerican.com/article.cfm?id=ocean-impact-map.

(Image 3.2.a. Pacific Ocean, Costa Rica)

2001 through 2010 was the warmest decade on record for both land and ocean surface temperatures. As one would expect, with increased temperatures comes increased melting of sea ice, glaciers, and ice sheets in Greenland and in the Antarctic, which, along with expansion of the water due to warming, has led to a rise in sea level. The average annual rise in sea level in the 20th century was 1.6 mm per year while the average rise in sea level in the first decade of the 21st century was 3 mm per year.[60,61]

The majority of the sea level rise has been due to melting of the Antarctic and Arctic ice, with artic sea ice decreasing 3 to 4 percent annually over the last three decades and large pieces of the West Antarctic shelf breaking off over the last two decades. A visual representation of the loss of Artic ice can be seen in Image 3.2.b. The Antarctic Ice Shelf is glacially fed while the Arctic ice is formed from the buildup of frozen sea water and precipitation. Since the Arctic ice is not glacially fed, increased temperatures mean that it will be difficult for the ice shelves to recover. The loss of sea ice not only

---

[60] World Meteorological Organization, *The Global Climate 2001-2010: A Decade of Climate Extremes - Summary Report* (Geneva: World Meteorological Organization, 2013), http://library.wmo.int/pmb_ged/wmo_1119_en.pdf.
[61] Thomas R. Karl, Jerry M. Melillo, and Thomas C. Peterson (eds.), *Global Climate Change Impacts in the United States* (New York: Cambridge University Press, 2009).

increases sea level, but it increases the warming of ocean water because sea ice reflects sunlight while open water absorbs sunlight, and it affects ocean currents, humidity, and cloudiness.[62,63]

(Image 3.2.b. Decline in Perennial Sea Ice between 1980 and 2012[64])

---

[62] Derek R. Mueller, Warwick F. Vincent, and Martin O. Jeffries, "Break-up of the Largest Arctic Ice Shelf and Associated Loss of an Epishelf Lake," *Geophysical Research Letters* 30, 20 (2003) 1-4.
[63] Thomas R. Karl, Jerry M. Melillo, and Thomas C. Peterson (eds.), *Global Climate Change Impacts in the United States* (New York: Cambridge University Press, 2009).
[64] NASA website, accessed February 17, 2017, https://www.nasa.gov/topics/earth/features/thick-melt.html.

With the melting of the Arctic ice and the loss of the Western Antarctic ice sheet, scientists expect sea level to rise to four meters over the next several centuries. If this expectation holds true, it could displace millions of people around the world. Specifically, in the United States, a third of all residents live in counties immediately bordering the coastline. If sea level were to rise by two feet, expectations for sea level rise in major cities include 2.3 feet in New York City, 2.9 feet in Hampton Roads, Virginia, and 3.5 fee in Galveston, Texas. Other effects of sea level rise will include loss of habitat for islands and inundation of low-lying areas around the world which could greatly impact food production as croplands in the Middle East and Asia could disappear.[65,66]

Another impact of climate change is an increase in the frequency and intensity of hurricanes. According to Gutowski et al (2008), "It is very likely that the human-induced increase in greenhouse gases has contributed to the increase in sea surface temperatures in the hurricane formation regions. Over the past 50 years there has been a strong statistical connection between tropical Atlantic sea surface temperatures and Atlantic hurricane activity."[67] As indicated in a report by the U.S. National Oceanic and Atmosphere Administration, 2001 to 2010 was the most active decade for hurricane activity in the North Atlantic since 1955.[68] In addition, some of the most intense and destructive hurricanes on record have occurred recently with 53% of recorded category 5 Pacific hurricanes occurring since 2002, and 87% occurring since 1994, and with 29% of recorded category 5 Atlantic hurricanes occurring since 2003.[69,70]

According to Karl et al (2009)[71], climate change is leading to changes in the water cycle including the following:
- changes in precipitation patterns and intensity
- changes in the incidence of drought
- widespread melting of snow and ice
- increasing atmospheric water vapor
- increasing evaporation
- increasing water temperatures
- reductions in lake and river ice
- changes in soil moisture and runoff

---

[65] Suzanne Goldenberg, "West Antarctic ice collapse 'could drown Middle East and Asia crops,'" *The Guardian* (May, 22 2014), https://www.theguardian.com/environment/2014/may/22/west-antarctic-ice-collapse-middle-east-asia-crops.

[66] Thomas R. Karl, Jerry M. Melillo, and Thomas C. Peterson (eds.), *Global Climate Change Impacts in the United States* (New York: Cambridge University Press, 2009).

[67] William J. Gutowski, Jr., Gabriele C. Hegerl, Greg J. Holland, Thomas R. Knutson, Linda O. Mearns, Ronald J. Stouffer, Peter J. Webster, Michael F. Wehner, and Francis W. Zwiers, "Causes of Observed Changes in Extremes and Projections of Future Changes" in *Weather and Climate Extremes in a Changing Climate. Regions of Focus: North America, Hawaii, Caribbean, and U.S. Pacific Islands*, ed. Thomas R. Karl, Gerald A. Meehl, Christopher D. Miller, Susan J. Hassol, Anne M. Waple, and William L. Murray (A Report by the U.S. Climate Change Science Program and the Subcommittee on Global Change Research, Washington, DC, 2008), p 81-116.

[68] Eric S. Blake and Ethan J. Gibney, *The Deadliest, costliest, and most intense United States tropical cyclones from 1851 to 2010 (and other frequently requested hurricane facts)*, (Miami: NOAA Technical Memorandum NWS NHC-6, 2011), 49 pp.

[69] "List of Category 5 Atlantic Hurricanes," Wikipedia, accessed February 17, 2017, https://en.wikipedia.org/wiki/List_of_Category_5_Atlantic_hurricanes.

[70] "List of Category 5 Pacific Hurricanes," Wikipedia, accessed February 17, 2017, https://en.wikipedia.org/wiki/List_of_Category_5_Pacific_hurricanes.

[71] Thomas R. Karl, Jerry M. Melillo, and Thomas C. Peterson (eds.), *Global Climate Change Impacts in the United States* (New York: Cambridge University Press, 2009).

Precipitation has increased by 5% over the past 50 years with incidents of heavy precipitation increasing approximately 20% during the same timeframe (see Image 3.2.c). 2001-2010 was the second wettest decade since 1901 with 2010 being the wettest year on record. Scientists anticipate that future precipitation events will be less frequent but more intense. For example, there has been little change or even a decrease in the frequency of light and moderate precipitation during the past 30 years, while heavy precipitation has increased.[72,73,74]

---

[72] Thomas R. Karl, Jerry M. Melillo, and Thomas C. Peterson (eds.), *Global Climate Change Impacts in the United States* (New York: Cambridge University Press, 2009).

[73] William J. Gutowski, Jr., Gabriele C. Hegerl, Greg J. Holland, Thomas R. Knutson, Linda O. Mearns, Ronald J. Stouffer, Peter J. Webster, Michael F. Wehner, and Francis W. Zwiers, "Causes of Observed Changes in Extremes and Projections of Future Changes" in *Weather and Climate Extremes in a Changing Climate. Regions of Focus: North America, Hawaii, Caribbean, and U.S. Pacific Islands*, ed. Thomas R. Karl, Gerald A. Meehl, Christopher D. Miller, Susan J. Hassol, Anne M. Waple, and William L. Murray (A Report by the U.S. Climate Change Science Program and the Subcommittee on Global Change Research, Washington, DC, 2008), p 81-116.

[74] World Meteorological Organization, *The Global Climate 2001-2010: A Decade of Climate Extremes - Summary Report* (Geneva: World Meteorological Organization, 2013), http://library.wmo.int/pmb_ged/wmo_1119_en.pdf.

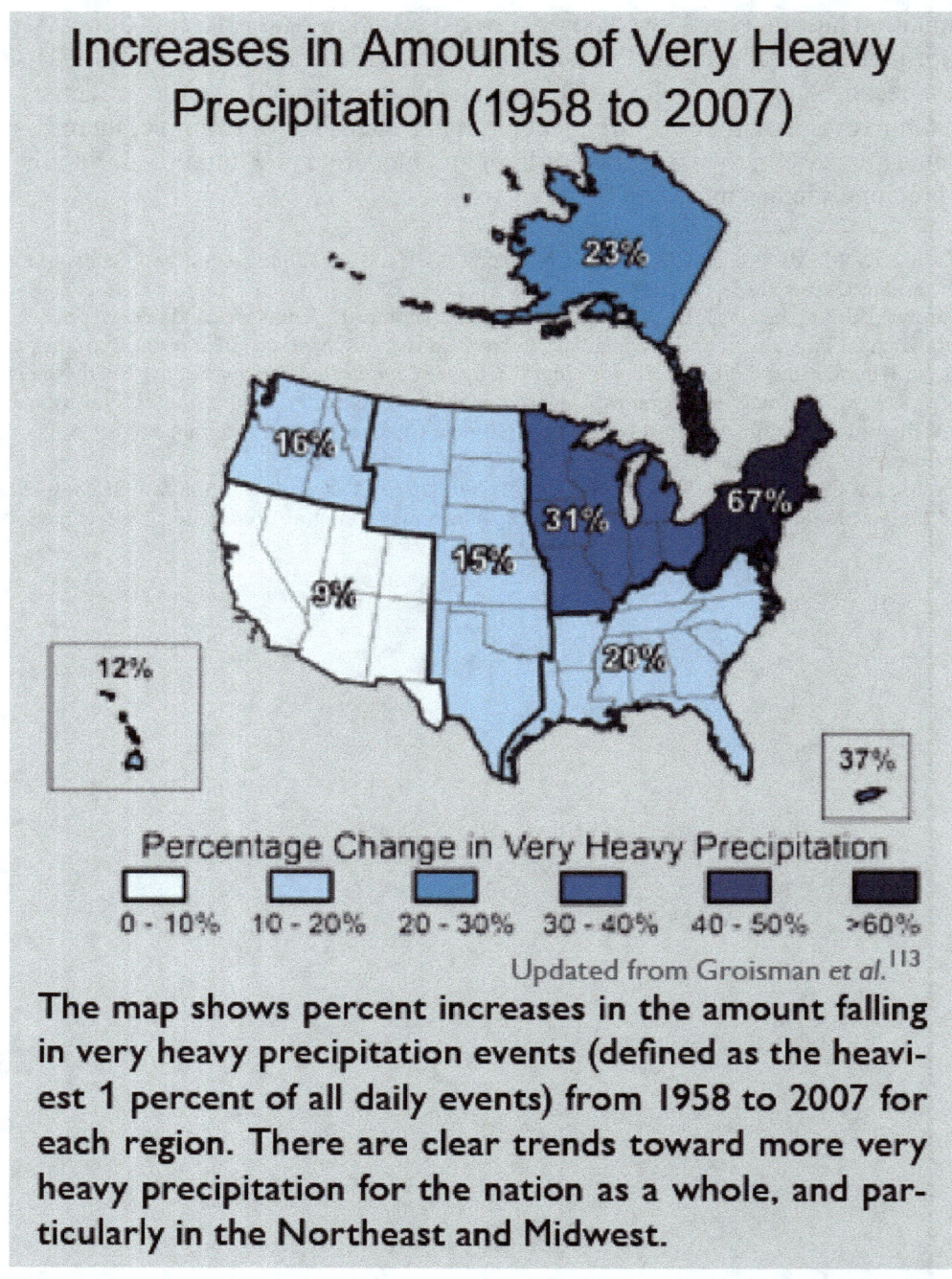

(Image 3.2.c. Increased heavy precipitation in the United States.[75])

An example in precipitation extremes can be seen in California. California entered a drought in 2012, with the Governor formerly declaring a drought in 2014. However, the winter of 2016 / 2017

---

[75] Thomas R. Karl, Jerry M. Melillo, and Thomas C. Peterson (eds.), *Global Climate Change Impacts in the United States* (New York: Cambridge University Press, 2009).

will likely break the drought and California is on track to have the wettest winter on record with the snowpack in the Sierra Mountains at the beginning of March at 185% of average.[76,77]

More intense precipitation events combined with less frequent precipitation has led to an increase in floods and droughts, with floods being the most frequent extreme weather event between 2001 and 2010. Examples of floods include flooding in Pakistan in 2010 where 2,000 people died and 20 million were affected) in 2010, flooding in Colorado in 2013 which led to 6 deaths, 1,500 homes destroyed, and 17,000 homes damaged, and flooding in California in 2017 where ~14,000 people were forced to evacuate their homes in San Jose after a week of heavy rain. In 200,8 the Mississippi River reached 7 feet above flood stage, which had effects on cropland including destroying many crops, causing serve erosion and runoff of surface chemicals into waterways, and leading to agricultural losses of $8 billion.[78,79,80,81]

---

[76] Kalhan Rosenblatt, "California on Track to Have Wettest Year on Record Following Five-Year Drought," NBC News, February 23, 2017, http://www.nbcnews.com/news/weather/california-track-have-wettest-year-record-following-five-year-drought-n724376?cid=eml_onsite.

[77] "Near-record Sierra snowpack — 185 percent of average," SFGATE, March 2, 2017, http://www.sfgate.com/bayarea/article/Near-record-Sierra-snowpack-185-percent-of-10969482.php.

[78] Thomas R. Karl, Jerry M. Melillo, and Thomas C. Peterson (eds.), *Global Climate Change Impacts in the United States* (New York: Cambridge University Press, 2009).

[79] Kalhan Rosenblatt, "California on Track to Have Wettest Year on Record Following Five-Year Drought," NBC News, February 23, 2017, http://www.nbcnews.com/news/weather/california-track-have-wettest-year-record-following-five-year-drought-n724376?cid=eml_onsite.

[80] World Meteorological Organization, *The Global Climate 2001-2010: A Decade of Climate Extremes - Summary Report* (Geneva: World Meteorological Organization, 2013), http://library.wmo.int/pmb_ged/wmo_1119_en.pdf.

[81] William J. Gutowski, Jr., Gabriele C. Hegerl, Greg J. Holland, Thomas R. Knutson, Linda O. Mearns, Ronald J. Stouffer, Peter J. Webster, Michael F. Wehner, and Francis W. Zwiers, "Causes of Observed Changes in Extremes and Projections of Future Changes" in *Weather and Climate Extremes in a Changing Climate. Regions of Focus: North America, Hawaii, Caribbean, and U.S. Pacific Islands*, ed. Thomas R. Karl, Gerald A. Meehl, Christopher D. Miller, Susan J. Hassol, Anne M. Waple, and William L. Murray (A Report by the U.S. Climate Change Science Program and the Subcommittee on Global Change Research, Washington, DC, 2008), p 81-116.

(Image 3.2.d. Iguazu Falls, Argentina)

      Drought is also expected to increase with climate change. According to a World Meteorological Organization Report in 2013, "Droughts affect more people than any other kind of natural hazards owing to their large scale and long-lasting nature." Specifically, increased temperatures increase the potential for evaporation, leading to increased drought. This will affect crops as they will require additional water with increased temperature, and soil will dry out. The trickle-down effects to the economy and growing population could be monumental.[82,83,84]

---

[82] World Meteorological Organization, *The Global Climate 2001-2010: A Decade of Climate Extremes - Summary Report* (Geneva: World Meteorological Organization, 2013), http://library.wmo.int/pmb_ged/wmo_1119_en.pdf.

[83] Thomas R. Karl, Jerry M. Melillo, and Thomas C. Peterson (eds.), *Global Climate Change Impacts in the United States* (New York: Cambridge University Press, 2009).

[84] William J. Gutowski, Jr., Gabriele C. Hegerl, Greg J. Holland, Thomas R. Knutson, Linda O. Mearns, Ronald J. Stouffer, Peter J. Webster, Michael F. Wehner, and Francis W. Zwiers, "Causes of Observed Changes in Extremes and Projections of Future Changes" in *Weather and Climate Extremes in a Changing Climate. Regions of Focus: North America, Hawaii, Caribbean, and U.S. Pacific Islands*, ed. Thomas R. Karl, Gerald A. Meehl, Christopher D. Miller, Susan J. Hassol, Anne M. Waple, and William L. Murray (A Report by the U.S. Climate Change Science Program and the Subcommittee on Global Change Research, Washington, DC, 2008), p 81-116.

Suggestions to reduce carbon dioxide emissions listed in Section 3, Day 1, will help reduce many of the topics discussed in this chapter including ice melt and sea level rise, intense storms, and flooding. Some suggestions to reduce water usage and, in turn, help reduce droughts are as follows:[85]

1. Take shorter showers.
2. Turn off the sink water when brushing your teeth.
3. Plant drought resistant grass and plants.
4. Insulate pipes to get hot water faster.
5. Check pipes, toilets, and hoses for leaks.

Additional water conservation tips can be found at the Water Use It Wisely website: http://wateruseitwisely.com/100-ways-to-conserve/

---

[85] "100+ Ways to Conserve Water," Water Use it Wisely, accessed April 6, 2017, http://wateruseitwisely.com/100-ways-to-conserve/.

Questions from day two:

1. What does firmament mean?
2. Where is Heaven in this three-part design of creation?
3. What is happening to the earth that is causing concern for scientists?
4. If the earth continues to warm, what will be the result?
5. What can we all do to prevent this warming?

Day Three: Gathering of the Water and Dry Land

[9] And God said, "Let the waters under the heavens be gathered together into one place, and let the dry land appear." And it was so.
[10] God called the dry land Earth, and the waters that were gathered together he called Seas. And God saw that it was good.
[11] And God said, "Let the earth put forth vegetation, plants yielding seed, and fruit trees bearing fruit in which is their seed, each according to its kind, upon the earth." And it was so.
[12] The earth brought forth vegetation, plants yielding seed according to their own kinds, and trees bearing fruit in which is their seed, each according to its kind. And God saw that it was good.
[13] And there was evening and there was morning, a third day.

In this third act of creation, two things occur. The first is the gathering together of the waters. This referred to the waters under the heavens not the waters above and below the heavens. By gathering together, the waters, the dry land that must have been under water, now appears.

The scripture tells us that God named these things that were created. The dry land was called "Earth" and the water was called "seas".

We know that there are three ways in which mountains are formed. These are known as volcanic, fold and block mountains. All of these are the result of plate tectonics, where compressional forces, isostatic uplift and intrusion of igneous matter forces surface rock upward, creating a landform higher than the surrounding features. We are aware of vast mountain ranges on the east and west side of the United States. Geologists love to study the rock formation that tells them how old it is and was stacked layer upon layer on top of each section. The vegetation or plant-life that was covered produces coal, gas, diamonds and in some cases, leaves behind fossils of plants and animals no longer living today.

Over the course of many million years, these uplifted sections are eroded by the elements – wind, rain, ice and gravity. These gradually wear the surface of the mountains down, cause the surface to be younger than the rocks that form them, and lead to the types of formations and distributions we are familiar with today.

Verse eleven is the second part of the creation that occurs on this day. Now that there was land, God declares that the earth should put forth vegetation, and it does. The earth produces these plants. Plants bear seeds and fruit trees bear fruit that have seeds. It appears that the author is making a point that the plants that were created were those that reproduce and continue to be useful for the rest of creation.

On this third day, we have dry land appear and seed-bearing plants. Plants are an important part of our world, and a major component of human nutrition. Humans may eat plants directly, may eat fruit or vegetables produced by plants, or may eat animals that rely on plants for their food. In addition, plants are major components of flavors, drinks, dyes, and products such as soap, clothing, and plastics, they decay into fossil fuels such as coal, and are often major components in pharmaceuticals.

Another important fact about plants is that most of them conduct photosynthesis – a process by which chlorophyll containing plants use light to convert carbon dioxide into energy, and release oxygen into the atmosphere. The increased burning of fossil fuels and deforestation is leading to an increase in carbon dioxide in the atmosphere as discussed in Section Three, Day One.

With the impact of plants in reducing atmospheric carbon through photosynthesis, it is important to understand worldwide state of plants. The current estimate for plant species in the world is ranges from 300,000 to ~500,000, with the most reliable estimates being in the 350,000 range. More than one in eight plant species in the world, and one in three in the United States, is under threat of extinction. Among the plants most at risk, are 14% of rose species, 14 % of cherry species, 29% of palm species, 32% of lily species, and 32% percent of iris species. Coniferous trees as a group are also seen as vulnerable. Additional details on threatened plants as discussed by Stevens (1998), are presented in see Figures 3.3.a and 3.3.b.[86]

Country	Number of Threatened Plant Species
United States	4,669
Australia	2,245
South Africa	2,215
Turkey	1,876
Mexico	1,593

(Figure 3.3.a. Number of threatened plant species by country.)

Country	Threatened Plant Species (% of Total Species)
United States	29.0
Jamaica	22.5
Turkey	21.7
Spain	19.5
Australia	14.4

(Figure 3.3.b. Percentage of threatened plant species by country.)

Over the past 8,000 years, nearly half of the Earth's forests have been converted to farms, pastures, and other uses. Figure 3.3.c details changes to the Earth's forests from their original state through 1997. Categories include cleared forest (forests that have been removed), frontier forest (original forests), and non-frontier forest (forests that had been cleared at one point that have since regrown).

---

[86] William K. Stevens, "Plant Survey Reveals Many Species Threatened With Extinction," *New York Times*, April 9, 1998, A-1.

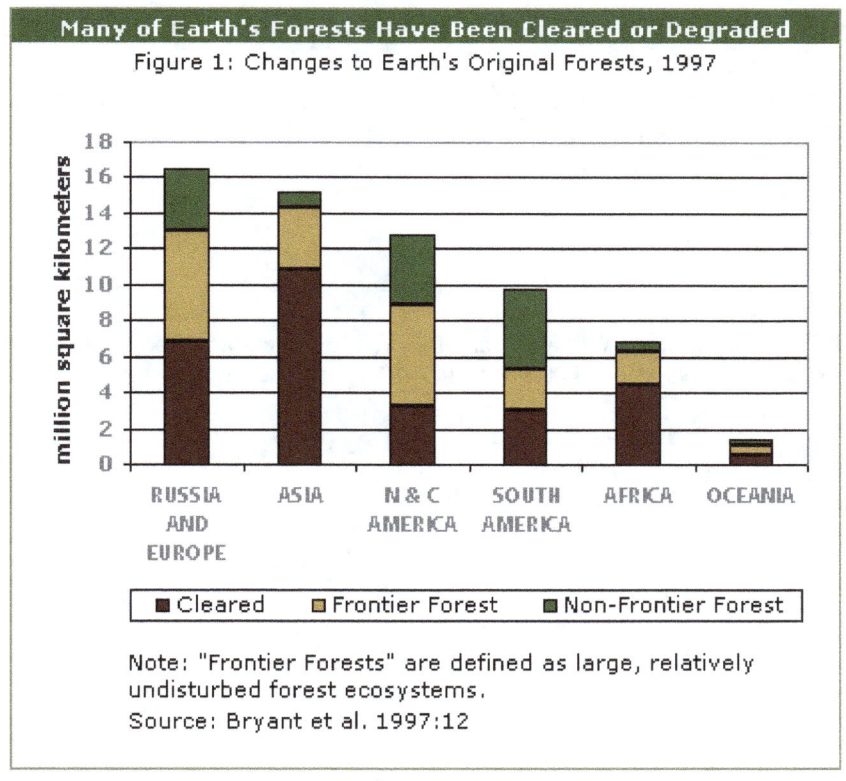

(Figure 3.3.c. Changes to Earth's original forests.[87])

But the human impact on forests did not stop with clearing large forest tracts as humans have heavily altered most of the remaining forests resulting in a patchwork of fragmented forests through a process referred to as forest fragmentation. According to a 1997 World Resources Institute assessment, just one fifth of the Earth's original forest remains in large, relatively natural ecosystems.[88]

Fragmentation caused by anthropogenic factors leads to changes in forest functions such as fire prevention and changes in ecological processes such as nutrient cycling, pollination, and decomposition.[89] For example, Mac Nally et al. (2000)[90] indicated that fragmentation driven by land clearing for agricultural use affects natural habitats by depleting the most fertile land. Agricultural use introduces chemicals, such as nitrogen used in fertilizers, into the environment. Additionally, processes such as wildfires and logging have detrimental effects on soils.[91] In cases where forest is

---

[87] Dirk Bryant, Daniel Nielsen and Laura Tangley, *The Last Frontier Forests: Ecosystems and Economies on the Edge*, (Washington, D.C.: World Resources Institute, 1997).

[88] Dirk Bryant, Daniel Nielsen and Laura Tangley, *The Last Frontier Forests: Ecosystems and Economies on the Edge*, (Washington, D.C.: World Resources Institute, 1997).

[89] Claude Gascon, Richard O. Bierregaard, Jr., William F. Laurance, and Judy Rankin-de Mérona, "Deforestation and Forest Fragmentation in the Amazon. In Lessons from Amazonia," in *The Ecology and Conservation of a Fragmented Forest,* ed. Richard O. Bierregaard, Jr., Claude Gascon, Thomas E. Lovejoy, and Rita Mesquita (Yale University Press, New Haven and London, 2001), 22-30.

[90] Ralph Mac Nally, Andrew F. Bennett and Gregory Horrocks, "Forecasting the Impacts of Habitat Fragmentation. Evaluation of Species-specific Predictions of the Impact of Habitat Fragmentation on Birds in the Box-ironbark Forests of Central Victoria, Australia," *Biological Conservation* 95 (2000): 7-29.

[91] J. Barlow, C.A. Peres, L.M.P. Henriques, P.C. Stouffer, and J.M. Wunderle, "The Responses of Understorey Birds to Forest Fragmentation, Logging and Wildfires: An Amazonian Synthesis," *Biological Conservation*, 128 (2006): 182-192.

fragmented due to road construction, species are introduced to many variables such as pollution, noise, road kill, human disturbance, exotic species, and reduced patch connectivity.[92]

We simply have not taken into account the animals and their habitat when we build new roads or cut down trees. Our concern is for our own convenience. This is not what is expected of us as caretakers of the creation.

---

[92] Rebecca A. Reed, Julia Johnson-Barnard, William L. and Baker, "Contribution of Roads to Forest Fragmentation in the Rocky Mountains," *Conservation Biology*, 10, 4 (1996): 1098-1106.

Suggestions on how to conserve plants and forests are as follows[93]:

1. Use double sided paper for printing.
2. Use paperless billing.
3. Use natural pesticides.
4. Read digital books.
5. Plant more trees and plants in your yard.

Additional ideas on how to conserve plants can be found at the Conserve Energy Future website: http://www.conserve-energy-future.com/fabulous-ways-to-protect-trees-and-conserve-forests.php

---

[93] "25+ Fabulous Ways to Protect Trees and Conserve Forests" Conserve Energy Future, accessed April 6, 2017, http://www.conserve-energy-future.com/fabulous-ways-to-protect-trees-and-conserve-forests.php.

Questions from day three:

1. How do you think God "Gathered the dry lands?"
2. Could our understanding of volcanic, plate and block mountains fit into what the Genesis account meant when it said that God "Created the dry land?"
3. If over 8,000 years, half of the earth's surface has been destroyed, what will be the effect of the rest of nature?
4. What are some of the results of fragmentation?

Day Four: Lights for Day and Night

[14] And God said, "Let there be lights in the firmament of the heavens to separate the day from the night; and let them be for signs and for seasons and for days and years,
[15] and let them be lights in the firmament of the heavens to give light upon the earth." And it was so.
[16] And God made the two great lights, the greater light to rule the day, and the lesser light to rule the night; he made the stars also.
[17] And God set them in the firmament of the heavens to give light upon the earth,
[18] to rule over the day and over the night, and to separate the light from the darkness. And God saw that it was good.
[19] And there was evening and there was morning, a fourth day.

The fourth day brings lights to the firmament in order to separate the day from the night. You may recall, however, that light had already been created on the first day. This may be an indication that there were multiple authors that edited the original P source. The redundancy certainly is inconsistent.

Once again, the Story Teller's Companion to the Bible refers to the firmament as the vault.

The purpose of the lights created on this day was to separate day from night. When the sun rises, day begins. When the moon shines through the darkness, night begins. And furthermore, the sun and the moon were to serve for "signs and for seasons and for days and for years". You may recall that the ancient Hebrews along with most of their neighbors interpreted what they saw in the sky at night as a sign from a divine being. The visit of the Magi at Jesus' birth is an example.

The author apparently knew enough about astronomy to understand that the sun stayed out longer in the summer months and that the moon shown in a cycle from full to slight each month. He did not know the earth was round and tilted on its axis as it rotated around the sun but knew that the moon and perhaps the stars and planets varied in their appearance.

To add additional confusion to what seems like a natural sequence, it appears that verses 16-19 were added by an editor to the original P source because they are redundant.[94] Verses 16a, 17a, 18a have a certain didactic intent which characterizes verse 14a. But here the author states the purpose for what was being created.

It is interesting to note the sequence of the plants being created before the sun. We know of course that the plants need sunlight to grow, but that does not seem to be understood by the author as he set forth his account of creation.

When we think of separation of day and night, we think of the sun (Image 3.4.a) and the moon. The sun has many impacts on the Earth. One topic not discussed in our earlier discussion, that ties in with this text, is that of photosynthesis. The right amount of light and heat from the sun is essential for nearly all life on Earth as light is a component in photosynthesis. The combination of water, carbon dioxide, and energy from sunlight lead to the production of sugars in the photosynthesis process. The sugars are used in autotrophs (most plants) for growth. Additionally, most heterotrophs (animals using plants as at least part of their diet) use the energy from sunlight through consumption of autotrophs, products of autotrophs, or heterotrophs that consume autotrophs. Heterotrophs subsequently break down the sugar products of photosynthesis providing their cells with energy. Thus, sunlight is involved in the production of energy for most life on Earth.

---

[94] Ibid. p.477

(Image 3.4.a. Sunset in Costa Rica)

The impact of the moon on Earth is subtler than that of the sun. We see it mostly through minimal light production and gravitational impacts. These gravitational force of the moon (and the sun) on the Earth leads to a "pulling" of the water on Earth toward the moon. Since the Earth is rotating, the pull is not constant, which leads to an ebb and flow, or high and low tides which typically once or twice daily along ocean coastlines. Twice per month, during a new moon and a full moon, the Earth, moon, and sun align, exerting a greater than normal gravitational pull on the Earth leading to spring tides (named for the tides spring forth), or higher than normal high tides. The moon does not maintain a constant distance from the Earth, with the moon being closest to the Earth every 3 to 4 months. With full or new moons when the moon is closest to the Earth, we see the greatest impact of the gravitational force on the tides with the highest high tides. With sea level rise discussed in Section 3, Day 2, the impact of high tides will be enhanced. When other weather impacts such as wind and storms are factored in, the higher than normal water levels experienced during high tides can lead to high surf and flooding.[95]

---

[95] Kevin Bryne, "What Effects Can the Full Moon Have on Weather, People and Animals?" Jul 23, 2016, https://www.accuweather.com/en/weather-news/do-full-moons-affect-weather-p/40127763.

Another impact of separation of day and night is that of nighttime cooling. Nighttime, with the absence of radiation from the sun, allows for our bodies, plants, animals, and infrastructure to shed heat and cool. However, increasing temperatures is increasing the number of hot days and nights. Ben Webster of *The Times* writes that a Met Office study found that the number of nights in which the temperature in cities stays above 68°F will increase fivefold because of climate change. At higher temperatures, the elderly and children become vulnerable to heat exhaustion. For example, a heat wave in Britain in 2003 in which nighttime temperatures were around 68°F and daytime temperatures were above 86°F for ten days in a row led to 2,000 deaths because people could not recover from the heat during the nighttime hours. With homes without air conditioning, the difference in temperature outside and inside the house at higher temperatures makes is difficult for people to escape from the heat. This effect is amplified in cities that act at head islands during the day with buildings and roads absorbing and holding heat, and then releasing the heat at night. Met Office suggests that by 2040, the government will issue heat wave warnings four times as frequently as they do today.[96]

A simple suggestion to increase photosynthesis would be to plant more plants. Other suggestions include:

1. Instead of grass, plant shrubs and trees that are indigenous to your area. They will grow, provide food and shelter to animals and reduce the amount of grass that needs to be cut, thereby reducing the carbon monoxide that comes from the mowers.
2. Plant a garden to grow some of your own food and help the environment.
3. Plant trees around your yard to provide shade to cool your house and a home for animals.

---

[96] B. Webster, "Night-time Temperatures Could Rise Above 25C Because of Climate Change," *The Times*, June 1, 2010, http://www.thetimes.co.uk/article/night-time-temperatures-could-rise-above-25c-because-of-climate-change-kwkztp5sm2m.

Questions related to day four:
1. What lights do we see in the firmament?
2. How do they help us?
3. How do the lights rule over the day and nights?
4. What ways do the sun and moon affect the earth?
5. How will continued climate change affect the way we exist when the days are very hot?

Day Five: Fish

[20] And God said, "Let the waters bring forth swarms of living creatures, and let birds fly above the earth across the firmament of the Heavens."
[21] So God created the great sea monsters and every living creature that moves, with which the waters swarm according to their kind.
[22] And God saw that it was good. And God blessed them saying, 'Be fruitful and multiply and fill the waters in the seas, and let birds multiply on the earth."
[23] And there was evening and there was morning, a fifth day.

It seems such a strange combination to create fish and birds on the same day, but that's what we have here. Perhaps, as some scholars suggest, the writer was hard pressed to get all of creation done in a week and so he doubles up here on fish and birds!

One of the most interesting things to note in these verses is that it wasn't just a pair of fish or birds of each species, there were, according to according to verse 21, " …all sea creatures with which the waters swarm…and every winged bird, according to its kind." There were a lot of fish and birds and God told them to multiply and God was pleased. In the telling, the ability to procreate occurred and those fish and birds present began to have more.[97]

This is the verse some Christians use to counter the evolution theories. They say that all birds and fish were created on this fifth day. In their minds, there is no need for evolution of the species.

Fish represent a food source for humans and animals, and in turn a variety of jobs are generated from activities associated with fish and seafood such as scuba and snorkeling, commercial and recreational fishing, food handling, food packaging, food delivery, and restaurants. In this section, we will look at the impact of climate change on fish, coral reefs, and bodies of water in general.

Increased air temperatures have been correlated with increased water temperatures in streams, rivers, lakes, and oceans. These increased temperatures are amplified in periods of low water flow in streams and rivers, and lead to longer periods of stratification (when bottom and surface waters do not mix) in lakes. Increased water temperature reduces dissolved oxygen content which affects aquatic animals and may lead to mortality. As discussed in Section 3, Day 2, heavy rain events lead to increased runoff of surface water. This runoff can carry pollution and pesticides into the waterways, lakes, and oceans, causing increases in algae, bacteria, and potentially leading to death in water species.[98,99]

Effects of climate change such as ocean warming, acidification and deoxygenation have historically contributed to mass extinctions of ocean dwelling animals. Bijma et. al. (2003)[100] suggest that the current rate of climate change combined with human impacts such as overfishing, eutrophication (increased nutrients in the water, primarily from runoff from the land) and pollution,

---

[97] Ibid p. 482
[98] Thomas R. Karl, Jerry M. Melillo, and Thomas C. Peterson (eds.), *Global Climate Change Impacts in the United States* (New York, 2009).
[99] M.C. Chiu, L. Hunt and V.H. Resh, "Climate-change Influences on the Response of Macroinvertebrate Communities to Pesticide Contamination in the Sacramento River, California Watershed," *Science of the Total Environment* 581-582 (2017): 741-749.
[100] J. Bijma, H.O. Portner, C. Yessen, A.D. Rogers, "Climate Change and the Oceans – What Does the Future Hold?," *Marine Pollution Bulletin* 74, issue 2 (2013): 495-505.

have never before been seen in history. Pitcher and Cheung (2013)[101] indicate that ocean production is likely to be reduced 2 to 13% by 2100 relative to 1860, and that the maximum body size of fish living in warmer waters will be reduced 14 to 24% by 2050 compared to current body sizes. The impacts of climate change also affect other seafood, such as shellfish, which are vulnerable to ocean acidification. Mollusks are an important food source in some countries such as Aruba, Turks and Caicos, and the Cook Islands, however, these fisheries are starting to decline since mollusks are susceptible to acidification. All of this taken in combination suggests that there will likely be fewer fish and other seafood, smaller fish, and less diverse ocean populations if current trends continue.

An area of significant concern is the impact of climate change on coral reefs. Some interesting facts about coral reefs are below:[102,103,104]

- Approximately 850 million people live within 60 miles of reefs, many of whom rely on reefs for food and associated jobs.
- More than 90,000 miles of shoreline in 100 countries and territories receive some protection from reefs, which reduce wave energy, erosion and storm damage.
- Reefs cover over 150,000 square miles of the ocean, which is less than 0.1% of the total marine environment.
- Reefs are home to at least 25% of known marine species, which includes 4,000 fish and 800 corals.
- Many reef dwelling organisms possess chemicals that are used in pharmaceuticals. As an example, a chemical originally found in sea squirts has been synthesized by Janssen and PharaMar for use as an anti-cancer drug named Yondelis to treat ovarian and other soft tissue cancers.

Juxtaposed with the importance of coral reefs to our society is the destruction of the reefs through human interactions and climate related changes. Figure 3.5.a details potential impacts of human interactions and climate change on coral reefs between the time that the image was published by Burke, 2011, and 2050. Human interactions such as overfishing, destructive fishing, marine and watershed pollution, and coastal development put over 60% of coral reefs in danger. In fact, overfishing and destructive fishing increased 80% from 1998 through 2011, with the largest increases seen in the Indian and Pacific Oceans. One driver for the increase is that populations living near reefs increased dramatically during that timeframe.[105]

When climate change such as changing water temperature and rising water levels are combined with the human interactions, the number of threatened coral reefs jumps to ~75%. The challenge with changing water temperatures is that coral reefs go through a process called coral bleaching, a process in which corals expel the algae living in them turning the corals white. Though the corals may continue to live, they are under more stress and are weaker. As an example, the Great Barrier Reef saw increased coral bleaching in 1998, 2002, and 2006, all years in which high water temperatures were recorded. Finally, circling back to Section 3, Day 2, increased hurricane activity and intensity

---

[101] Tony J. Pitcher and William W.L. Cheung, "Fisheries: Hope or Despair," *Marine Pollution Bulletin* 74, issue 2 (2013): 506-516.
[102] Daniel J. DeNoon, "Sea Squirt Drug Offers Cancer Hope." *WebMD*, June 21, 2007, http://www.webmd.com/cancer/news/20070621/sea-squirt-drug-offers-cancer-hope.
[103] Rhett A. Butler, "Coral Reefs Decimated by 2050, Great Barrier Reef's Coral 95% Dead." *Mongabay*, November 17, 2005. https://news.mongabay.com/2005/11/coral-reefs-decimated-by-2050-great-barrier-reefs-coral-95-dead/.
[104] Lauretta Burke, Katie Reytar, Mark Spalding, and Allison Perry, *Reefs at Risk Revisited*, (Washington D.C., World Resources Institute, 2011), http://www.wri.org/publication/reefs-risk-revisited.
[105] Lauretta Burke, Katie Reytar, Mark Spalding, and Allison Perry, *Reefs at Risk Revisited*, (Washington D.C., World Resources Institute, 2011), http://www.wri.org/publication/reefs-risk-revisited.

has been shown to lead to losses in biodiversity and potential extinction of reef dwelling species of plants and animals.[106,107,108]

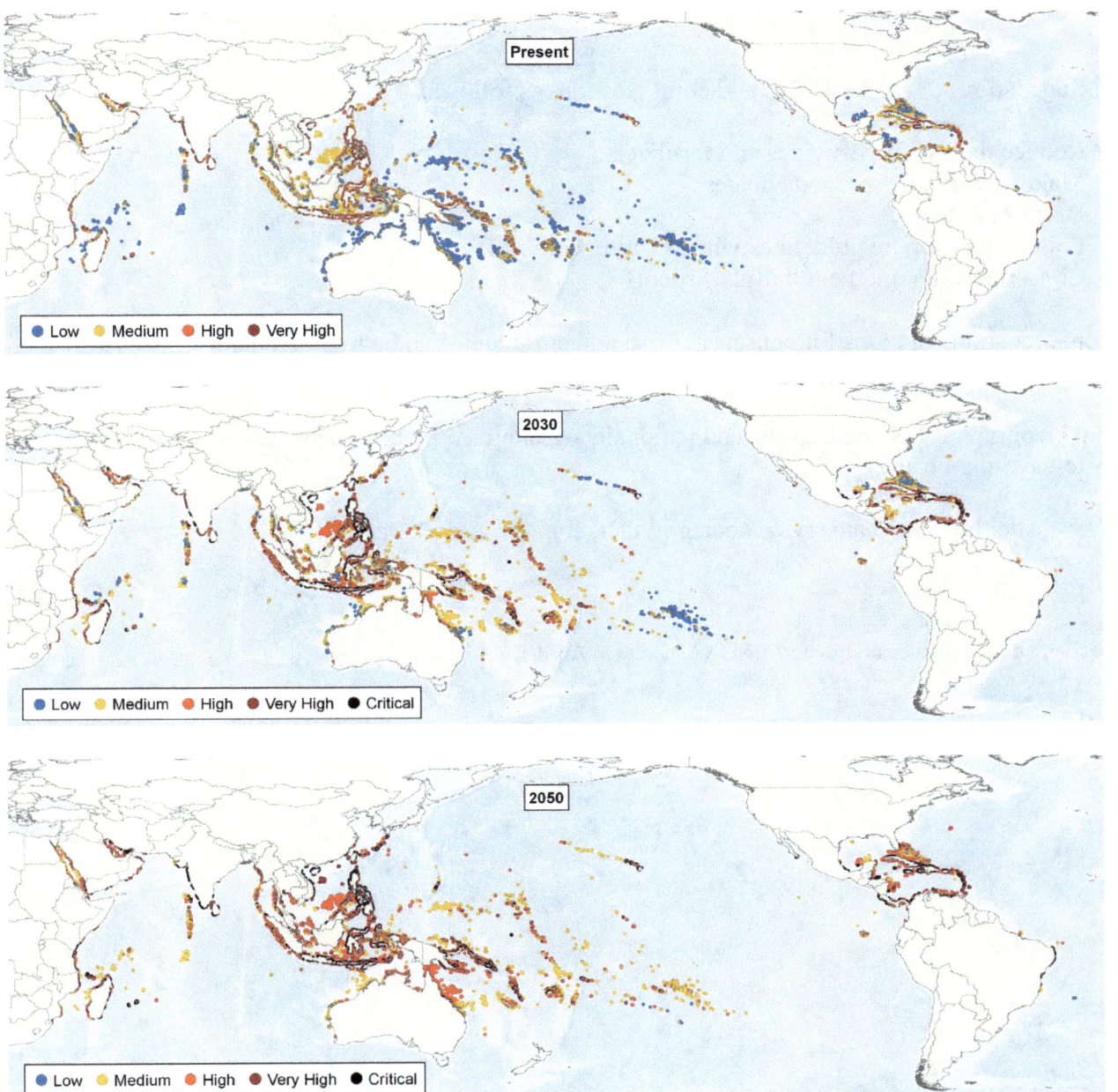

Image 3.5.a. Potential impacts of human interactions and climate change on coral reefs.[109]

---

[106] Lauretta Burke, Katie Reytar, Mark Spalding, and Allison Perry, *Reefs at Risk Revisited,* (Washington D.C., World Resources Institute, 2011), http://www.wri.org/publication/reefs-risk-revisited.
[107] "What is Coral Bleaching," NOAA, accessed March 29, 2017, http://oceanservice.noaa.gov/facts/coral_bleach.html.
[108] Jorge Christian Alva-Basurto and Jesus Ernesto Arias-Gonzalez, "Modelling the Effects of Climate Change on a Caribbean Coral Reef Food Web," *Ecological Modelling* 289 (2014): 1-14.
[109] Lauretta Burke, Katie Reytar, Mark Spalding, and Allison Perry, *Reefs at Risk Revisited,* (Washington D.C., World Resources Institute, 2011), http://www.wri.org/publication/reefs-risk-revisited.

Suggestions to conserve fish and coral reefs are as follows:[110]

1. Reduce the use of pesticides and fertilizers.
2. Make sustainable seafood choices.
3. Don't pollute.
4. Follow appropriate guidelines when visiting coral reefs.
5. Conserve water and be mindful of runoff.

Additional recommendations for conserving fish and coral reefs can be found at the following websites:

National Geographic website http://ocean.nationalgeographic.com/ocean/take-action/10-things-you-can-do-to-save-the-ocean/

NOAA website: https://oceanservice.noaa.gov/facts/thingsyoucando.html

---

[110] "What can I do to Protect Coral Reefs?," NOAA, accessed April 6, 2017, http://www.publicaffairs.noaa.gov/25list.html.

Questions to consider for day five:
1. Is it possible that saying that God created the fish and the birds could allow for the evolution of those fish and animals based on the influences also created by God?
2. What can we do to protect coral reefs?
3. What can we do to protect fish in streams, rivers, lakes and oceans?
4. What percent of all marine species live in reefs?
5. Human interactions such as overfishing, destructive fishing, marine and watershed pollution, and coastal development put what percent of coral reefs in danger?

Day Six

[24] "And God said, 'Let the earth bring forth living creatures according to their kinds; cattle and creeping things and beasts of the earth according to their kinds.' And so, it was."

[25] "God made the wild animals of the earth of every kind, and the cattle of every kind and everything that creeps upon the ground of every kind."

[26] "Then God said, 'Let us make humankind in our image, according to our likeness; and let them have dominion over the fish of the sea, and over the birds of the air, and over the cattle, and over every creeping thing that creeps upon the earth."

[27] So, God created humankind in his imagine, in the image of God he created them, male and female he created them.

[28] God blessed them, and God said to them, "Be fruitful and multiply, and fill the earth and subdue it; and have dominion over the fish of the seas and the birds of the air and over everything that moves upon the earth."

[29] God said, "See, I have given you every plant yielding seed that is upon the face of all the earth, and every tree with seed in its fruit; you shall have them for food."

[30] And to every beast of the earth, and to every bird of the air, and to everything that creeps on the earth, everything that has the breath of life, I have given you every green plant for food." And so, it was.

[31] God saw everything that he had made and, and indeed, it was very good. And there was evening and there was morning, the sixth day.

The animals are represented here as being produced by the earth; much like the plants had been in verse eleven.

The animals are divided into three groups. There are the domestic cattle, the creeping things; reptiles, incest and small quadrupeds and the beasts of the earth. We assume the beasts to be wild animals that are carnivore.

It is important to note here that there is no blessing of the animals similar to that of the fish and the birds. Scholars have offered various opinions as to why. The most plausible is that in the original narrative, only man was blessed and that the blessing of the fish and the birds was a reflex of verse 28.

The creation of man and woman is somewhat different from the other creations. For the first time in these verses, God uses the second person plural pronoun. That makes us wonder with whom is God speaking? Williams in "The Story Tellers Companion to the Bible", believes that "The structure of the plot gives us a clue to the context; look at what comes before and after the statement for a possible answer. On the previous day, God created the animals, both wild and domesticated. Is God then speaking to them, or perhaps, speaking about them in the second-person plural pronouns of verse 26."[111]

Since we know that this story came down through the years and that it was influenced by other cultures, it may be possible that this verse came from another creation account wherein there were multiple gods. The text tells us that God first consults with other divine beings. The words, "lets us make man in our image, after our likeness" leads us to believe this was the case. There is also the reference to the Spirit that was present when the world was void and without form. The use of "us" and "our" could simply be plural of majesty. The Babylonian creation epic that we provided in the first section of this book, includes an instance when Marduk, before creating man, declared his intention to Ea. Knowing for sure is impossible. But simply noting that the difference exists in our scripture is important, not because it makes us of greater worth, but because of the greater responsibility that will

---

[111] *The Story Tellers Companion to the Bible.* Michael E. Williams, editor. (Nashville, 1976) p. 27.

be given to us after this consultation. The verse continues "…and let them have dominion over…" listing all the things that had been created in verse 20-24. Being created in the image of God must surely mean more than looking like God. Very few of us look alike. Could this refer to the power of thought or communication or self, transcendence?[112]

Reading verse 27 points out one obvious difference from the second creation story in Genesis 2. Here both men and women are created at the same time. In Genesis 2, man is created first, then plants and animals and finally, women. But this follows the creation thyme of creating blessing plants, fish, birds and animals. In each case, God told these new creations to be fruitful and multiply.

Verse 29 is not good news for those of us who eat meat. This verse tells us that the plants were given to provide food for man and birds and animals. Somewhere the sheep, and goats were used to eat rather than just provide milk and wool. Somewhere along the way, the wild beasts got tired of eating grass and started eating smaller animals. Somewhere along the way, large fish started eating small fish and birds of prey started eating other birds and small animals. How all of this evolved over the centuries is unknown, but it is a certainty. To be faithful to the scriptures and the command from God, I suppose we need to eat more green and less red.

The theme of God's sovereignty over all creation comes to a climax with the creation of man and woman. By placing this act last in this account, the P writer depicts humanity as the crown of all creation. Man and woman were made in the image of God. They were to be a living representation of God to rule on earth. The nobility of humanity is that they are given a special task among all things created. They were to be fruitful and multiply and exercise dominion over the empire God created.

There was one restriction given humanity in this creation account. In verses 29 and 30 we find that they were to eat only fruits and vegetables, as alluded to above.

The priestly narrative of creation comes to a conclusion with the creation of the Sabbath, for according to this account, after six days God rested, which in Hebrew is shabbath. We find in Exodus 16 that as the centuries went by, the Hebrews established a day of rest and gladness, a day of Sabbath rest. The Hebrews began to use this day as a day to refrain from all work and to worship their God. The priestly writer was aware of these traditions and included the day of rest in this creation account.

Scientists estimate that nearly nine million eukaryotic species (all living organisms except bacteria, blue-green algae, and other primitive microorganisms) exist on earth. Of the 8.74 million estimated species, approximately 7.77 million are animals. Climate change is affecting and will affect species, including plants and animals, in various ways. Climate changes will shift biomes, or communities of living organisms in a major ecological region. For example, warmer temperatures will allow species to move farther up mountain peaks, raising the tree line and in turn raising the habitat for animals in the biome (Image 3.6.a.). These changes will lead to reduction of alpine and subalpine habitats, potentially leading to isolation of species living in those regions. In addition, warmer weather will lead to many species shifting farther from the equator and closer to the poles. Species shifting closer to the poles may decrease the ranges and available resources to those animals already living in the area.[113,114,115,116]

---

[112] *The Interpreter's Bible*, (Nashville,1952) Volume I, p. 485.
[113] Census of Marine Life. "How Many Species on Earth? About 8.7 Million, New Estimate Says." ScienceDaily, accessed April 5, 2017, www.sciencedaily.com/releases/2011/08/110823180459.htm
[114] "Climate Change - Effects on Animals, Birdlife and Plants," Climate and Weather, accessed April 5, 2017, http://www.climateandweather.net/global-warming/climate-change-and-animals.html.
[115] Michael J. Case, Joshua J. Lawler and Jorge A. Tomasevic, "Relative Sensitivity to Climate Change of Species in Northwestern North America," *Biological Conservation* 187 (2015): 127-133.

(Image 3.6.a. Grand Teton Mountains in Jackson Hole, Wyoming.)

Climate change will be a benefit to species that are easily able to adapt and expand their ranges including pests and cold-sensitive species such as the Burmese python in Florida.[117] However, the speed at which these changes are occurring is limiting the ability of most species to adapt. According to Krause and Farina, "complex interactions triggered by the extreme events associated with climate change have a strong impact inside trophic chains with cascade effects, like the irreversible transformation of habitats, the rapid extinction of species, and dramatic changes in entire communities where key species are involved."[118]

Some effects of climate change on animals are as follows:
- With shorter winters and warmer weather, mammals come out of hibernation sooner.
- A sea level rise of 20 inches could cause sea turtles to lose their nesting beaches.

---

[116] Susan Joy Hassol, *Impacts of a Warming Arctic: Arctic Climate Impact Assessment*, (Cambridge, Cambridge University Press, 2004), 140 pp.

[117] Christine Dell'Amore, "7 Species Hit Hard by Climate Change—Including One That's Already Extinct," *National Geographic*, April 2, 2014, http://news.nationalgeographic.com/news/2014/03/140331-global-warming-climate-change-ipcc-animals-science-environment/.

[118] Bernie Krause and Almo Farina, "Using Ecoacoustic Methods to Survey the Impacts of Climate Change on Biodiversity," *Biological Conservation* 195 (2016): 245-254.

- Temperature changes can affect gender differences in reptiles and amphibians. For example, a 3°F increase in temperature will lead to a higher number of female sea turtles in southern United States.
- The distribution of crickets in Ecuador is highly dependent on temperature. A change in temperature in the lowland forests in which they live could alter the distribution and survival of various species.
- Many amphibians depend on seasonal wetlands and streams, which will be altered by warming and changes in rainfall.
- Reductions in sea ice will drastically shrink marine habitat for polar bears, ice-inhabiting seals, and some seabirds, pushing some species toward extinction.
- Polar bears use Artic sea ice to hunt for food, but the ice is melting earlier in the summer and forming later in the fall than normal. The decline in sea ice will lead to polar bears needing to adapt their hunting styles to hunt on land for survival, or we may see the extinction of the species.
- Caribou and reindeer in the Artic are likely to be stressed as climate change alters their access to food, breeding grounds, and migration routes.[119,120,121,122,123]

A 2017 publication by the Medical Society Consortium on Climate and Human Health titled "Medical Alert! Climate Change is Harming our Health"[124] starts with a powerful first paragraph:
"Most Americans understand that climate change is real and are concerned about it. But most still see climate change as a faraway threat, in both time and place, and as something that threatens the future of polar bears but not necessarily people. The reality, however, is starkly different: climate change is already causing problems in communities in every region of our nation, and from a doctor's perspective, it's harming our health."

The authors break the impacts of climate change on human health into three main categories: Direct Health Harms, Spreading Disease, and Disrupting our Well Being. Image 3.6.b is a visual representation of climate related health affects in the US.

---

[119] "Climate Change - Effects on Animals, Birdlife and Plants," Climate and Weather, accessed April 5, 2017, http://www.climateandweather.net/global-warming/climate-change-and-animals.html.

[120] Bernie Krause and Almo Farina, "Using Ecoacoustic Methods to Survey the Impacts of Climate Change on Biodiversity," *Biological Conservation* 195 (2016): 245-254.

[121] Michael J. Case, Joshua J. Lawler and Jorge A. Tomasevic, "Relative Sensitivity to Climate Change of Species in Northwestern North America," *Biological Conservation* 187 (2015): 127-133.

[122] Susan Joy Hassol, *Impacts of a Warming Arctic: Arctic Climate Impact Assessment*, (Cambridge, Cambridge University Press, 2004), 140 pp.

[123] Christine Dell'Amore, "7 Species Hit Hard by Climate Change—Including One That's Already Extinct," *National Geographic*, April 2, 2014, http://news.nationalgeographic.com/news/2014/03/140331-global-warming-climate-change-ipcc-animals-science-environment/.

[124] M. Sarfaty, R.J. Gould, and E.W. Maibach, "Medical Alert! Climate Change Is Harming Our Health," *The Medical Society Consortium on Climate and Health*, accessed April 5, 2017, https://medsocietiesforclimatehealth.org/reports/medical-alert/.

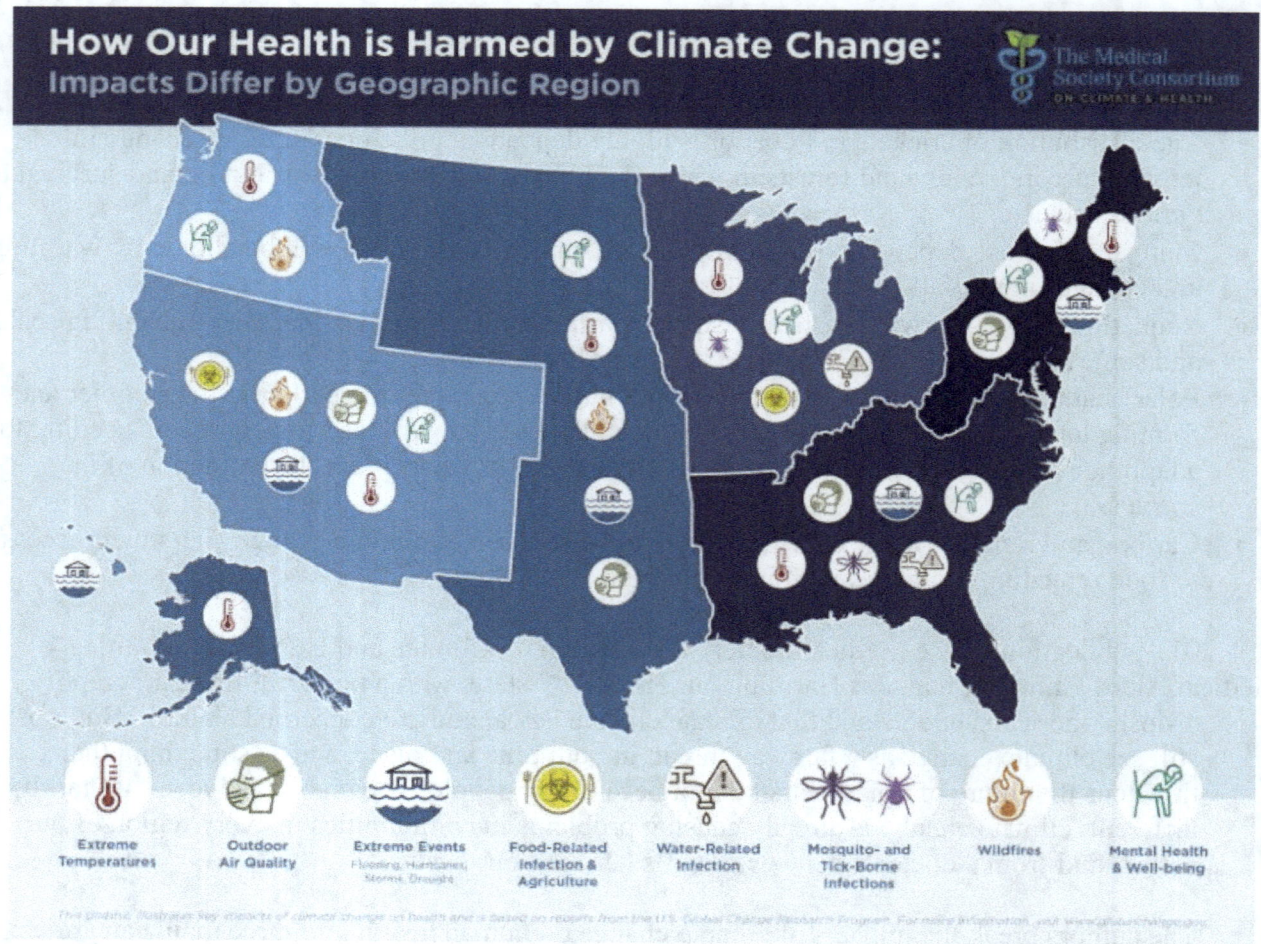

(Image 3.6.b. How our health is harmed by climate change. Medical Society Consortium on Climate and Health)

Direct Health Harms include extreme heat, extreme weather, and air pollution. We touched on extreme heat in Section 3, Day 4, with increase heat leading to hotter days and nights, and longer and more frequent heat waves. The authors suggest that increased heat can lead to death from heat stroke and dehydration, with the great affects being seen in athletes, people living in cities, those without air conditioning, those that work outside, the young and the elderly, and pregnant woman.

We touched on extreme weather in Section 3, Day 2, and the authors suggest that climate change is "causing increases in the frequency and severity of some extreme weather events such as heavy downpours, floods, droughts, and major storms." These changes can lead to displacement, injury, and death. They may reduce the ability of people to receive medical care or medications, may knock out power leading to increased dangers from heat or cold, and may lead to increased illnesses from waterborne or insect carried diseases.

The authors indicate that climate change is affecting air quality as increased heat increases wildfires and smoke from those wildfires, pollen production, and smog, while increased heavy downpours as discussed in Section 3, Day 2 can lead to increased flooding and mold growth. These reductions in air quality can lead to increased allergy and asthma attacks, and increased incidences of lung cancer and heart disease.

The Spreading Disease category includes Ticks and Mosquitoes, Contaminated Water, and Contaminated Food. Ticks and mosquitoes can carry diseases such as West Nile Virus, Malaria, and Lyme disease. However, environmental conditions need to be right for ticks and mosquitoes to survive. Environmental changes due to climate change such as increased heavy downpours have led to increased standing water in some areas, which is a situation in which mosquitoes thrive. Lyme disease carrying ticks are expanding their range northward and westward due to warmer conditions, with Lyme disease being reported in 46% of US counties in 2017 versus only 30% in 1998.

Higher water temperatures, rising ocean levels, and increased heavy downpours as describe in Section 3, Day 2, are all leading to contamination of drinking water, recreational water, and seafood. Runoff of fertilizers into waterways combined with higher water temperatures leads to increased viruses, bacteria, parasites, and algal blooms, all of which can be harmful to humans. Rising temperature and flooding both lead to increases in contaminated food as floods can spread bacteria into fields from farms or other waterways, and pests, bacteria, and parasites thrive in higher temperatures.

In the final category, Disrupting our Well Being, the authors list Threats to Mental Health. Exposure to extreme weather such as storms, heat, flooding and drought, can lead to anxiety and depression. These disasters have also been associated with alcohol and drug abuse. With increased threat of stronger hurricanes come an increased chance of property damage. The anxiety and depression associated with losing one's house would fall into this category.

Suggestions to conserve animals are as follows[125,126]:
1. Plant native plants as they attract native species.
2. Reduce pesticide use.
3. Do not purchase products made from endangered or threatened species.
4. Visit national parks or reserves as these areas protect animals.
5. Add stickers to windows to prevent bird strikes.

Additional reading can be found at the following websites:

Endangered Species Coalition website: http://www.endangered.org/10-easy-things-you-can-do-to-save-endangered-species/

Conserve-Energy Future website: http://www.conserve-energy-future.com/30-astounding-ways-to-protect-and-conserve-wildlife.php

---

[125] "10 Easy Things you can do to Save Endangered Species," *Endangered Species Coalition*, accessed April 6, 2017, http://www.endangered.org/10-easy-things-you-can-do-to-save-endangered-species/.
[126] "30 Astounding Ways to Protect and Conserve Wildlife," *Conserve-Energy Future*, accessed April 6, 2017, http://www.conserve-energy-future.com/30-astounding-ways-to-protect-and-conserve-wildlife.php.

Questions to consider from day six:
1. What does it mean to "Have dominion over the fish, birds and animals?"
2. What responsibility do we have with our dominion?
3. How can we protect the creatures of the earth?
4. What can you do to reduce the harmful things we have been doing to our planet?

Section Four

Protecting God's Good Earth

The last verse in our biblical creation story regarding having dominion over the earth is a significant responsibility of humanity. It is not given to the beasts or the fish. It is given to humanity. This raises the question, "How are we following this first commandment from God?"

We live in a very complex world. Often, what benefits someone hurts another. Coal mining provided jobs for miners and fuel for steel mills which also provided jobs. But it also produced black lung disease for miners and smoke and smog filled cities where steel mills were located. The street lights burned day and night due to the smoke. These were dangerous jobs that provided a living for those willing to work in the mills and mines.

Strip mining was less dangerous for those who worked in them, but it was devastating to the mountains that were dug into and left bare of trees for animal life.

We are not always aware of the dangers we face in our work environments. Everyday there are advertisements on the television seeking people who have worked with asbestos. A law firm is suing a wide variety of companies who used asbestos. It was used in tile floors and ceilings, in automobile brakes, in insclation around hot water pipes and much more.

Farmers used chemicals to keep insects from eating their fruits and vegetables, but the DDT and other products they sprayed got into the water system as it ran off the fields and into the streams.

The concluding chapter of the book the *Environmental Debate* shares some information we believe to be helpful as we look back at what has been done to protect our environment. Prior to the early 1960's, there was little done to bring awareness to the attention of the politicians regarding the destruction of the environment. Between 1960 and 1979, however, we experienced the "Heyday of the Environmental Movement."[127]

In the early 60's agricultural and industry were at an all-time high in the United States. The increase in the production of goods and services between 1950 and 1970 matched the increase that had occurred between 1620 when the Pilgrims landed and 1950. That was tremendous growth, but there was a downside. That was an enormous demand for energy and a huge increase in oil imports. In the end, the growth led to consumption of huge amounts of natural resources and widespread industrial pollution.[128]

It was in Richard Nixon's presidency that the federal government finally began to commit itself to action to improve and protect the environment. He signed the National Environmental Policy Act in 1970 and the first Earth Day was created.

The National Environmental Policy Act created the Environmental Protection Agency which pushed for more and more improvements in the environment. It was during the 1980's that strong environmental legislation was passed by every level of government. There was the clean air act to stop pollution and the clean water act to prevent factories from dumping chemicals into the rivers. The clean water act also served to protect the wetlands, and essential part of our ecosystem, scenic areas, wildlife inhabitants, forests, marine mammals, endangered species, and human health. They also

---

[127] *The Environmental Debate*, second Edition, A Documentary History, with Timeline, Glossary and Appendices, edited by Peninah Neimark and Peter Rhodes Mott. P.190
[128] Ibid. p. 190

controlled the use of chemicals used by farmers as alluded to earlier, as well as radioactive and toxic wastes.[129]

It was a struggle because the rich and powerful who owned factories, mines and farms all wanted to produce more of their product cheaper, often at the cost of the health or lives of their employees.

In March of 2005, the United Nations published its Millennium Ecosystem Assessment, which was a very thorough assessment by over 1,000 scientists all over the world on the state of living systems on earth and how they are impacted by human activities. It begins by making it clear that "human activity is putting such strain on the natural functions of earth that the ability of the planet's ecosystems to sustain future generations can no longer be taken for granted."[130]

Part of the explanations given by John Houghton who writes this chapter is that we are using up more "Natural Capital" at an alarming rate and "living on borrowed time."[131] Houghton gives us an example and tells us that we are using up supplies of fresh ground water faster that it can be recharged and that many of the world's fish stocks are in a "dire state."[132]

The Millennium Assessment concluded with the simple message that "pressure on ecosystems will increase globally in coming decades unless human activities and actions change.[133]

Houghton challenges us to think about what help we can offer the world as it struggles with the major problems of our time. He suggests a theology of stewardship to form a solid foundation for our thinking about what God would have us do to save this creation.[134]

An article by David Fahrenthold in The Washington Post Magazine, presents a look at the Chesapeake Bay as an example of the many signs of human intrusion into the environment. He says that our cows are fouling a country stream, our nets are trapping too many fish, our homes and bright-blue swimming pools are colonizing a neck of bayside land. As the bay watershed spreads, we have spread with it and tried to make it work for us- as a seafood pantry, as a real-estate amenity, as a playground and as a gutter.

The bay has been around for more than 10,000 years, formed when melting glaciers drowned a long stretch of the Susquehanna River Valley. The Chesapeake watershed stretches over 64,000 square miles- its arms are big rivers such as the Susquehanna, the Potomac and the James, and its spider web fingers extend into sweaty tidal swamps out to the hard-rock folds of Appalachia and up the cold woods near Cooperstown, NY.

For centuries, these waters mixed with Atlantic Ocean backwash to make a brackish estuary and an ecological superconductor. The Chesapeake was alive with crabs, sturgeon and rockfish. Oysters grew on oysters, in reefs so bug they broke the surface of the water. The bay's bounty, which helped sustain Native Americans for centuries, seemed endless.

Then: the rest of us.

After Europeans arrived, their axes took out the trees that served as a natural water filter. Their ships scrapped away the bay's oysters. Their plows disturbed the earth, and then rain-water carried the earth downstream. Bladensburg was originally a deep-water port, but by 1830 the Anacostia River was too heavy with silt for big ships.

---

[129] Ibid. p. 192.
[130] Environmental Stewardship. Edited by R.J. Berry. T&T Clark International, London, New York. 2006 p. 315
[131] Ibid . P. 315.
[132] Ibid. p. 315.
[133] Ibid. p.316.
[134] Ibid. p.317.

Today's Chesapeake serves as a kind of living memory of those sins, and a few that we're still cooking up. More than 16.5 million people inhabit the Chesapeake watershed. Fertilizer from our lawns, manure from our farms and treated sewage from our cities help create oxygen-starved "dead zones" downstream in the bay. Toxic compounds from urban areas have been blamed for tumors on fish in the Anacostia and South rivers. And mysterious (but assuredly man-made) pollutants are creating other problems: Male bass in the Potomac are producing eggs and fish in the Shenandoah die in droves every spring.[135]

Bill Mc Kibben is a teacher, author and activist. He has written numerous books on the environment. We will use two of them to provide information regarding climate change.

He says that "NASA's James Hansen, the planet's premier climate change scientist, provided us with a number: in January 2008, his team published a paper showing that if the concentration of carbon dioxide in the atmosphere rose above 350 parts per million, we couldn't have a planet similar to the one which civilization developed and to which life on earth is adapted."[136]

Across the United States, in the five decades I've been alive, the number of extreme downpours a year has increased 30%. Across New England, it's gone up 85%. And all of this is with one degree of temperature increase.[137]

We have heard warnings from scientists for years and have been slow to make the recommended changes. Mc Kibben says, "The same climatologists who told us that this would happen now tell us we're likely to see temperatures rise five degrees this century unless we get off coal and gas and oil far faster than any government now plans."[138]

Each degree warmer the earth becomes brings with it more devastation. Mr. Kibben says, "If we move from one degree above average to four or five, that would be enough to turn farm belts into deserts, make New England a swamp and reduce Iceland to just another link in an endless chain of disasters that will turn civilization into a never-ending emergency response drill."[139] We have not heeded the warnings and many climatologists now insist that even if we stopped burning fossil fuels tomorrow, the temperature and the damages would increase for decades to come.

Mr. Mc Kibben quoted from the premier scientific journal, *Nature,* that published a new paper authored by twenty-three high profile biologists and climatologists, warning that "we were on the edge of a planetary 'state shift' that would leave the earth remarkably different than the one every human had here to fore known."[140]

We would like to share a few examples that Mc Kibben provides regarding what is happening as a result of climate change: "It wasn't just the 2,132 new high temperatures marks in June of 2012. It was what went with them. Duluth, Minnesota broke all its rainfall records, and in an excellent cinematic touch, so much water flooded the city zoo that the seal escaped and swam down Grand Avenue. In the Gulf of Mexico, meanwhile, tropical storm Debbie became the earliest fourth storm of the season ever recorded, and then dumped 'unthinkable amounts of rain' on central Florida."[141]

---

[135] By David Fahrenthold The Washington Post Magazine, June 7th 2009 fahrenthold@washpost.com.
[136] Bill McKibben, Oil and Honey, Henry Holt & Company LLC , New York, New York 2013 p 12
[137] Ibid. p.46.
[138] Ibid. p.40.
[139] Ibid. p 41.
[140] Ibid. p.140.
[141] Ibid. p.162.

"Out west, the largest fire in New Mexico history torched more than 171,000 acres, and then the most destructive blaze in the annals of Colorado burned on the edge of Fort Collins."[142]

A second book by McKibben offers even more details. We would like to share some of them here. "So far humans, by burning fossil fuel, have raised the temperature of the planet nearly a degree Celsius (since 1968). A NASA study in December 2008 found that warming on that scale was enough to trigger a 45% increase in thunder-heads above the ocean, breeding the spectacular anvil-headed clouds that can rise five miles above the sea generating 'super cells' with torrents of rain and hail. In fact, total global rain fall is now increasing 1.5% per decade."[143]

Besides Al Gore who sounded the alarm before any outside of the scientific world was ready to hear it, let alone heed it, President Obama in a speech at the University of New Hampshire said, "This is our generation's moment to save the future generations from global catastrophe."[144] And one of his political counterparts agreed. John McCain said a few months later, "We and the other nations of the world must get serious about substantially reducing greenhouse gas emissions in the coming years or we will hand off a much-diminished world to our grandchildren."[145] Influential scientists from around the world study and report on their findings. For many years, it was believed that the safe number at most, is 350 parts of carbon dioxide per million, but if we produce a carbon dioxide level above that, "we threaten the ecological life-support systems that have developed in the late Quaternary environment, and surely challenge the viability of contemporary human societies."[146] Over the past two hundred years, we have burned wood, coal, gas and oil that emitted carbon into the atmosphere and at the same time, we destroyed the tress that had been absorbing the carbon monoxide. That raised the temperature enough to start the process in motion. We have lived through the industrial revolution and now know that one barrel of oil yields as much energy as twenty-five thousand hours of human manual labor-more than a decade of human labor per barrel. The average American uses twenty-five barrels each year, which is like finding three hundred years of free labor annually.

Mc Kibben tells us that, "The planet we inhabit has a finite number of huge physical features. Virtually all of them seem to be changing rapidly: the Arctic ice cap is melting, and the great glacier above Greenland is thinning, both disconcerting and unexpected speed. The oceans which cover three fourths of the earth's surface, are distinctly more acid and their level is rising; they are also warmer, which means that the greatest storms on the planet, hurricanes and cyclones, have become more powerful. The vast inland glaciers in the Andes and Himalayas, and the giant snow pack of the American West, are melting very fast, and within decades, the supply of water to the billions of people living downstream may dwindle. The great rainforest of the Amazon is drying on its margins and threatened at its core."[147]

So, what's the good news? Well, there is some very good news in December of 2015, with the Paris Agreement at the 21st Conference of Parties (COP21). This is an international agreement that includes almost every country in the world. It focuses on reducing greenhouse gases, increasing adaptation and climate resilience, as well as climate finance.
It was negotiated and adopted by 195 countries. The agreement limited global temperature increase to 2 degrees Celsius, urging all to try to limit the emissions that cause increase to 1.5 degree Celsius. It requires all countries to make nationally determined contributions. It expects that all countries will

---

[142] Ibid. p.162.
[143] Bill McKibben, Earth, Time Books Henry Holt and Company, LLC. 175 Fifth Avenue, New York, New York, 10010 p. 3.
[144] Ibid. p.11.
[145] Ibid. p.11.
[146] Ibid. p.16.
[147] Ibid. p 49.

increase their commitments every five years and asks developed countries to help the developing countries. The final goal was to mobilize $100 billion a year to support mitigation and adaption.

The United Methodist Church publishes a *Book of Discipline* every four years after our General Conference. In the 2012 version, in the section called the "Social Principals", is found several paragraphs regarding the environment.

"All creation is the Lord's, and we are responsible for the ways in which we use and abuse it. Water, air, soil, minerals, energy resources, plants, animal life, and space are to be valued and conserved because they are God's creation and not solely because they are useful to human beings. God has granted us stewardship of creation. We should meet these stewardship duties through acts of loving care and respect. Economic, political, social and technological developments have increased our human numbers, and lengthened and enriched our lives. However, these developments have led to regional defoliation, dramatic extinction of species, massive human suffering, overpopulation, and overconsumption of natural and nonrenewable resources, particularly by individual societies. This continued course of action jeopardizes the natural heritage that God has entrusted to all generation. Therefore, let us recognize the responsibility of the church and its members to place a high priority on changes in economic, political, social, and technological lifestyles to support a more ecological equitable world leading to a higher quality of life for all of God's creation.

A) *Water, Air, Soil, Minerals, Plants*- We support and encourage social policies that serve to reduce and control the creation of industrial byproducts that serve and waste; facilitate the safe processing of disposal of toxic and nuclear waste and move forward the elimination of both; encourage reduction in municipal waste; provide for appropriate recycling and disposal of municipal waste; and assist the cleanup of polluted air, water, and soil. We call for the preservation of old-growth forests and other irreplaceable natural treasures, as well as preservation of endangered plant species. We support measures designed to maintain and restore natural ecosystems. We support policies that develop alternatives to chemicals used for growing, processing, and preserving food, and we strongly urge research into their effects upon God's creation prior to utilization. We urge development of international agreements concerning equitable utilization of the world's resources form human benefit so long as the integrity of the earth is maintained.

B) *Energy, Resources, Utilization*- The whole earth is God's good creation and as such has inherent value. We are aware that the current utilization of energy resources threatens this creation at its very foundation. As members of the United Methodist Church, we are committed to approaching creation, energy, production, and especially creation's resources in a responsible, careful and economic way. We call upon all to take measures to save energy. Everyone should adapt his or her lifestyle to the average consumption of energy that respects the limits of the planet earth. We encourage to limit CO2 emissions toward the goal of one tonne per person annually. We strongly advocate for the priority of the development of renewable energies. The deposits of carbon, oil and gas resources are limited, and their continuous utilization accelerates global warming"[148]

With this as our goal, the church continues to find ways to reverse climate change. At a Caretakers of God's Creation Climate Justice Conference in Washington, DC, on April 28, 2017, United Methodists held a day of prayer, worship and information gathering from scientists who study

---

[148] *The Book of Discipline of the United Methodist Church 2012*, Neil Alexander, publisher and editor, (Nashville, 2012), p.105

climate change. The next day, they joined thousands of other people from around the world to march to make their voices heard regarding their belief that we are not doing enough to save the planet and reverse a negative climate change.

At that conference, Dr. John Venezia, a principal at ICF International who has had 17 years' experience with greenhouse gas, made a presentation. He was kind enough to share his slides, some of which I will use to help make the point that climate change is real and that there are ways to combat it. He began by saying:

#1. Carbon in the atmosphere held at more than 400 parts per million for the first time...Carbon content of 360 parts per million ...is considered the trigger point for climate disruption;

#2. The death of 100 million trees in California is blamed on a five-year drought...and a subsequent attack by the pine bark beetle;

#3. The U.S. Fish and Wildlife Service proposed listing the rusty patched bumble bee, a native North American pollinator ...as an endangered species;

#4. Tufted puffins were dying in a part of the Bering Sea where scientists suspect...warming currents have shifted the food web. Thousands of snow geese died during the autumn migration when bad weather forced them to land at Berkeley Pit, a flooded [and severely contaminated] former copper mine in Montana;

#5. Since 2009 the bird-friendly Great Plains, which stretches from Texas to Canada, has lost 53 million acres of grasslands...;

#6. The world's population of birds has declined by hundreds of millions compared with just a few years ago because of climate disruption, dwindling habitat, hunting and pollution. The deaths of 80,000 reindeer in Russia were being tied to the retreat of Arctic sea ice...;

#7. Deforestation in the Amazon River basin was 29 percent higher than in 2015; and,

#8. Satellite images revealed Great Salt Lake in Utah is drying up, five years of drought having reduced the lake's surface by 40 percent..."

He then shared these news reports from, newspapers and magazines:

* Then this from the Times on March 19th, "...the Great Barrier Reef is dying...There is no mystery about the reason--it's global warming, caused by fossil fuels we burn...The death of coral reefs is a tragedy on many levels. There is the sheer beauty...they support a quarter of all marine life and provide protein for millions of people..."

* The Dispatch, again, in a telling front page story about Ohio beekeepers, who "reported a 44 percent colony loss during 2015. The stakes are high. Ohio farmers rely on bees to pollinate about 70 crops...It's estimated that bees play a role in the production of one-third of the food in the United States."

* And yet, this from the Times on March 17th, "Before he became President, Donald J. Trump called climate change a hoax...and mocked renewable energy as a plaything of 'tree-huggers.' So perhaps it is no surprise that Mr. Trump's first budget took direct aim at basic scientific and medical research...The White House is also proposing to eliminate climate science programs...including the Environmental Protection Agency...the director of the Office of Management and Budget said...'We

consider that to be a waste of your money...' The budget would eliminate money to carry out...plan[s] to reduce carbon dioxide emissions from coal-fired plants..."

*The Times further stated, "Industry tells Trump which rules to overturn" as EPA chief [Pruitt] calls for 'exit' from Paris climate agreement..."

* And, The New Yorker opined on April 10th, "President Trump said that his order puts 'an end to the war on coal.' In reality, it is a declaration of war on the basic knowledge of the harm that burning coal, and other fossil fuels, can do. Indeed, it tells the government to ignore information. He chooses to cast such worries aside at ...Mar-a-Lago, even as that property sinks into the rising sea, a process that has begun and, by many scientific estimations, will result in its grounds becoming one with the Atlantic during Barron Trump's lifetime."

Clearly, committed Caretakers of God's Creation have cause for despairing hearts and souls laden with at least a modicum of despair.

Al Gore, who served as Vice president under President Clinton, was absolutely correct when he wrote "along with the danger we face... this crisis also brings unprecedented opportunities... [of] a compelling moral purpose, a shared and unifying cause, the thrill of being forced by circumstances to put aside the pettiness and conflict that so often stifle the restless human need for transcendence, the opportunity to rise...Those who are now suffocating in cynicism and despair will be able to breathe freely. Those who are now suffering from a loss of meaning will find hope." [*An Inconvenient Truth, Introduction*]

The following slides were also from Dr. Venezia's presentation:

Much of my update on climate science and projections is from the latest Assessment Report from the Intergovernmental Panel on Climate Change, or IPCC.

The IPCC is a scientific and intergovernmental body under the United Nations, dedicated to the task of providing the world with an objective, scientific view of climate change and its political and economic impacts.

Christopher S. and Timothy M. Farabaugh

Emissions from the widespread burning of fossil fuels since the start of the Industrial Revolution have increased the concentration of greenhouse gases in the atmosphere.
And because these gases can remain in the atmosphere for hundreds of years before being removed by natural processes, their warming influence is projected to persist well into the next century.

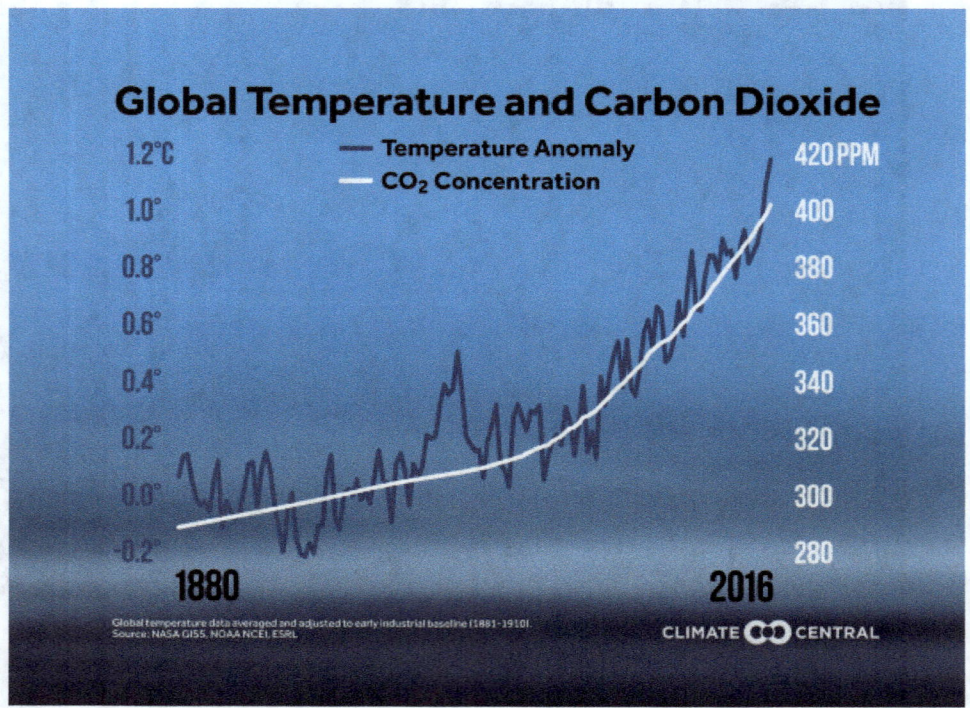

The IPCC finds that it is extremely likely that we are the dominant cause of warming since the mid-20th century. The overwhelming consensus of scientific studies on climate indicates that most of the observed increase in global average temperatures since the latter part of the 20th century is very likely due to human activities, primarily from increases in greenhouse gas concentrations resulting from the burning of fossil fuels.

As the concentration of carbon dioxide increases each year, so do temperatures. In fact, they correlate very well. Before the industrial revolution, the concentration of carbon dioxide in the atmosphere was around 280 parts per million. In the year 2016, we've now reached 400 parts per million.

Global annual average temperature has increased by more than 1.5°F since 1880.

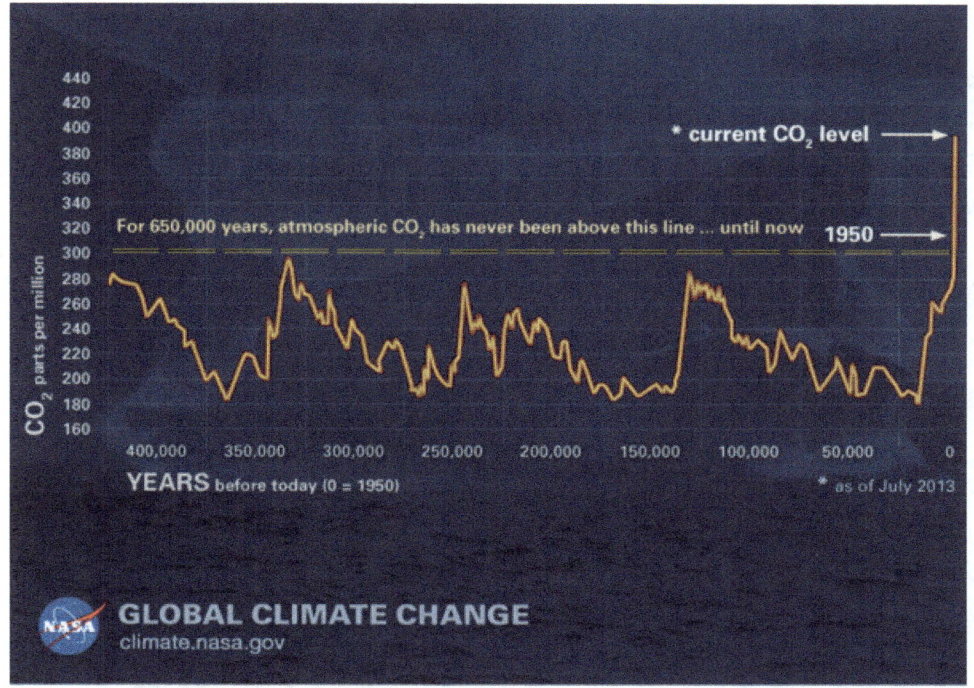

This chart from NASA shows historical atmospheric CO2 concentrations going back 400,000 years.

For 650,000 years, the concentration in the atmosphere has never been above 300 parts per million. So, 400 parts per million is uncharted territory, at least as long as humans have been on the earth.

The last time we were at 400 parts per million was 3 to 5 million years ago. Sea levels were anywhere from 15 to 130 feet higher than they are today. The earth was a very different place.

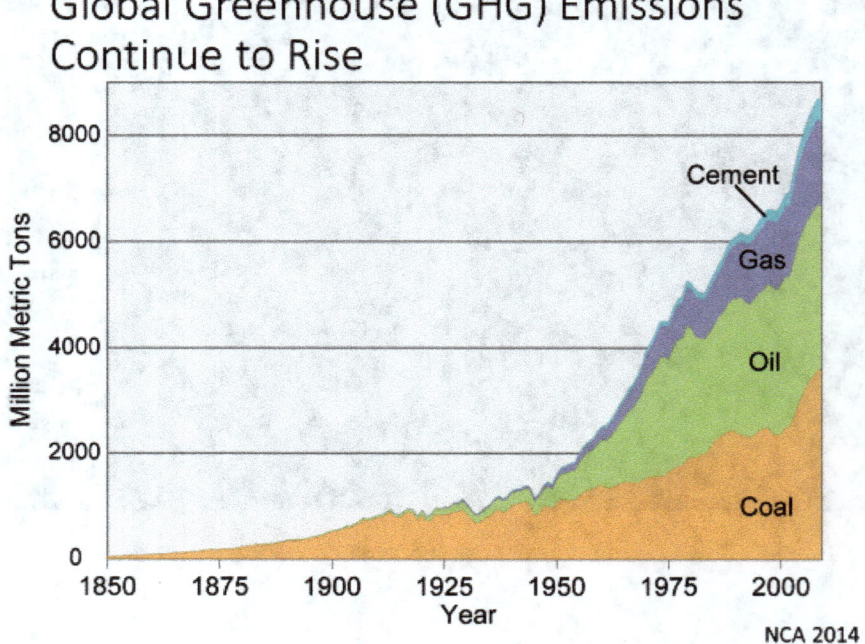

Carbon and other GHG emissions are emitted from burning of fossil fuels, largely for energy. These emissions account for about 80% of the total emissions of carbon from human activities, with land-use changes (like cutting down forests) accounting for the other 20%.

Even scarier, 16 of the top 17 hottest years have occurred since the year 2000

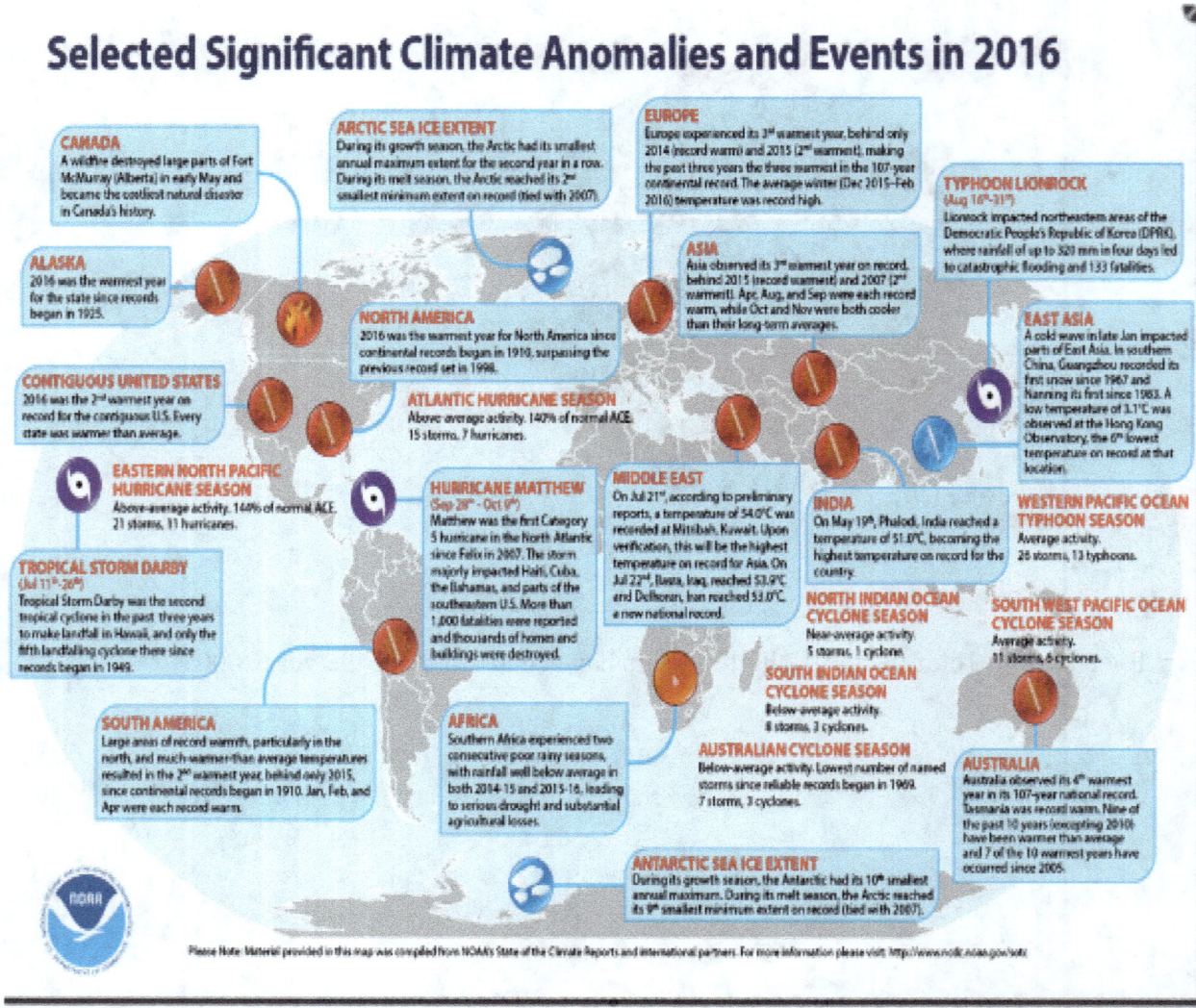

Increasingly, weather events around the globe can now be linked to climate change. This slide from the National Oceanic and Atmospheric Administration, or NOAA, shows just a few of the significant climate and weather events the world experienced in 2016.

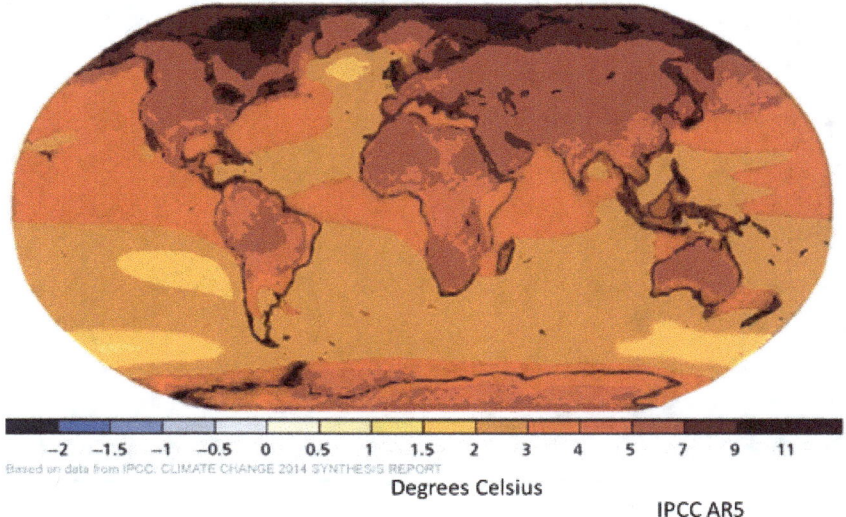

In a world where we do nothing to limit greenhouse gases, we expect temperatures to increase about 3 to 5 degrees Celsius by the year 2100, which translates to about 5 to 9 degrees Fahrenheit. And that's just a global average. Some places will experience much higher increases.

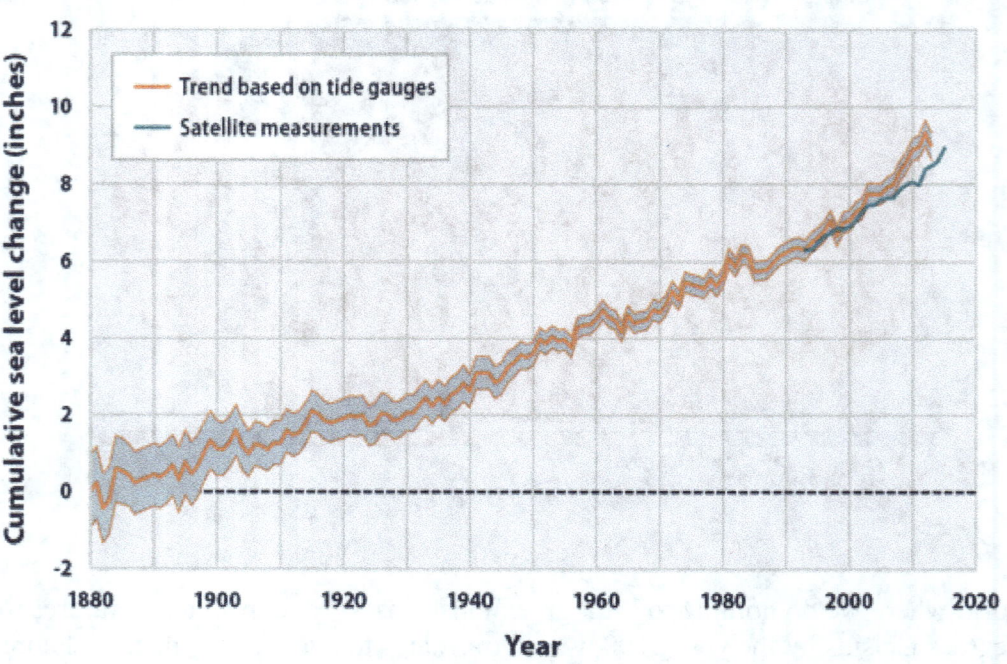

Global sea levels have risen, due to ocean expansion as they get warmer, as well as ice sheet melting.

Since 1900, average global sea levels have risen about 10 inches, which might not sound too bad.

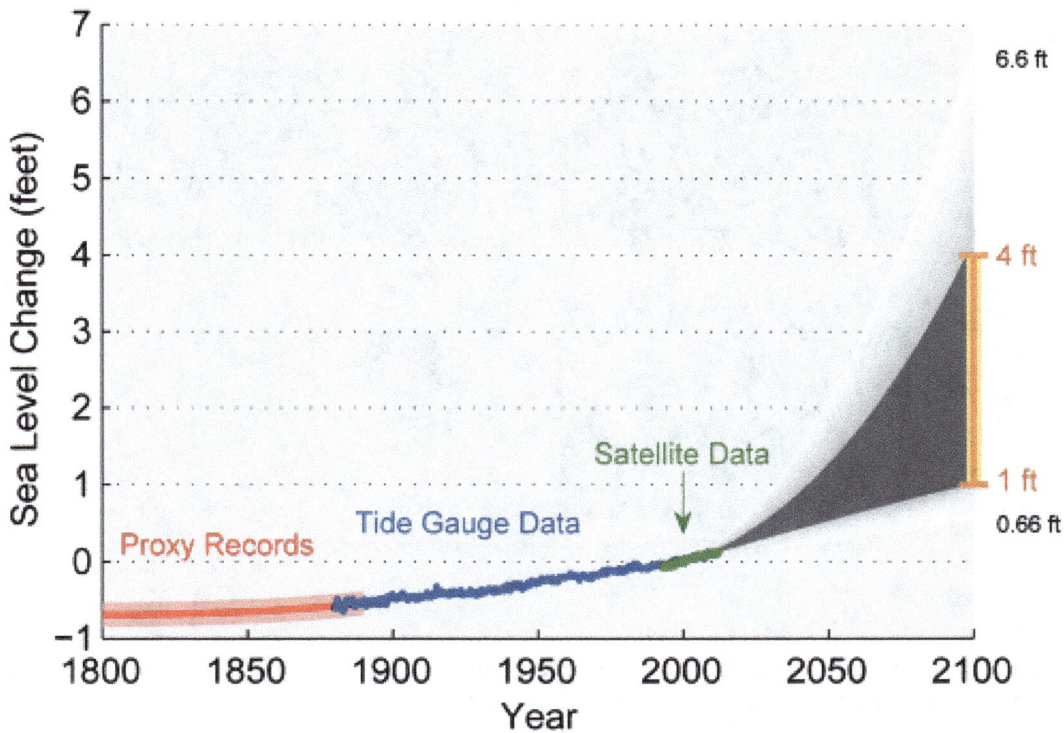

Figure source: NASA Jet Propulsion Laboratory

However, sea levels will continue to rise. NASA projects that by the end of this century, they will rise anywhere from 1 to 4 feet, with the possibility of increases of more than 6 feet. This would be devastating, as much of the world's population lives along the coastlines.

Credits: ©Scott Olson/Getty Images, NOAA, Alessandro Grassani, NCA 2014

We're looking at increasing droughts, severe coastal flooding, millions of climate refugees, food and water shortages, species loss, damages to infrastructure, more allergies, and much, much more.

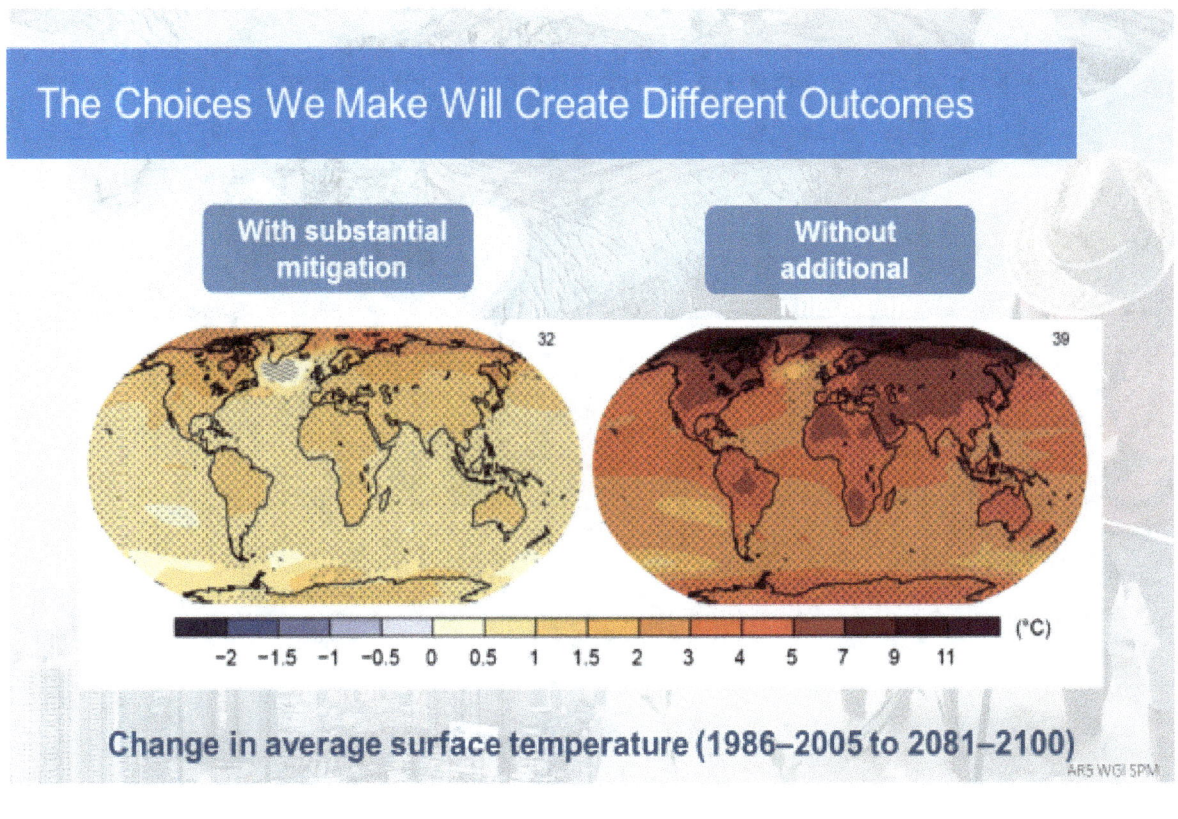

The choices we make today will have substantially differently outcomes for future generations. You can see here the projections of temperature increases with GHG mitigation on the left, as opposed to the much warmer temperatures on right, assuming we don't act.

Paris Agreement is Important, But Not the Only Answer

While the Paris Agreement is important, it's not the only game in town. In fact, on June 1, 2017, President Trump decided to pull out of the Paris agreement to the dismay of hundreds of countries. He had the support of 22 GOP Senators, who received $10 Million from the fossil fuel industry. They wrote a letter to Trump, urging him to pull out of the Paris Climate Agreement... and he did! But at that news, governors, mayors and CEO's from national corporations all pledged their support for the Parris agreement and promised to continue to support it. The Paris agreement is just at the national level.

States will continue to take action. 20 U.S. states still have ambitious GHG targets. Many major U.S. cities, and corporations are staying committed to reducing emissions.

Cities and provinces around the world have organized into groups such as the Global Covenant of Mayors for Climate and Energy, C40, and Under 2 MOU, under which they commit to reducing their emissions.

But clearly, we need to keep up the pressure on our local representatives, state representatives, as well as this Administration.

In addition, we need to stay committed at home to reducing emissions in our daily lives.

Companies are Increasingly Making Climate and Renewable Energy Commitments

- 50% of Fortune 500 companies have a climate or clean energy target
- 63% of Fortune 100 companies have a clean energy target
- Many have 100% renewable targets

Source: World Wildlife Fund

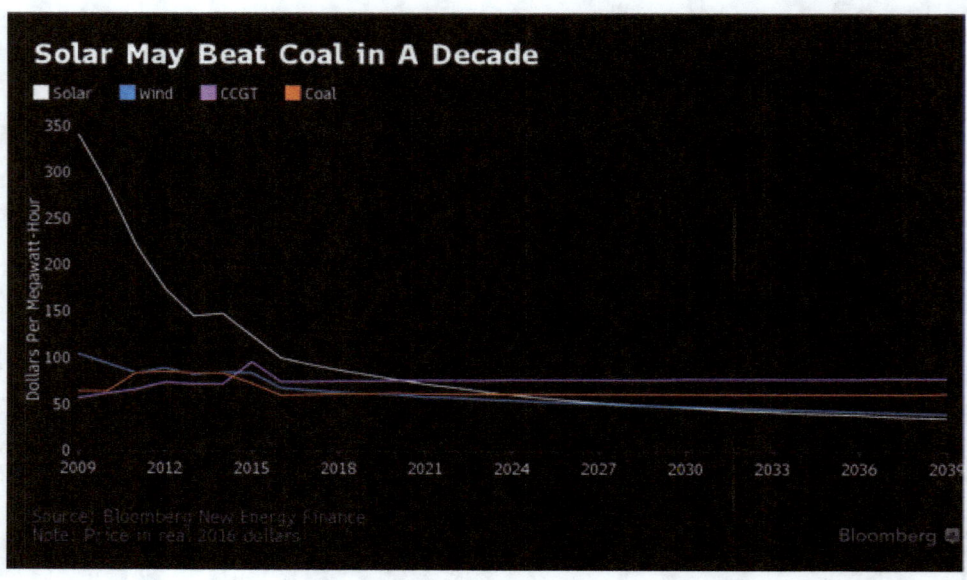

Another thing that's promising is economics. The cost of wind, and especially solar in recent years has plummeted. It depends highly on the country and location, but in many countries solar and wind are actually now, cheaper than coal. The United States Navy has just installed a field of solar panels at NAS Oceana in Virginia Beach, VA. The Navy partnered with Dominion Virginia Power to build a 21-megawatt solar facility on the Navy's East Coast master jet base. That's enough energy to power about 2,100 homes in Virginia, according to the Solar Energy Industry Association.

Under the agreement, Dominion will build, own, operate and maintain the Oceana facility for 37 years. In exchange, Oceana will receive electrical infrastructure upgrades.

"These projects increase the energy security, energy diversity and energy resiliency of our bases. Energy security or having assured access to reliable supplies of energy and the ability to protect and deliver sufficient energy to meet mission-essential requirements, is critical to our installations' roles to support the Fleet," Rear Adm. Jack Scorby, commander of Navy Region Mid-Atlantic, said in a statement.

Reducing reliance on fossil fuels has long been one of Navy Secretary Ray Mabus' priorities in office. The Navy has set a goal of getting 50 percent of its total energy consumption from alternative sources by 2020. The Navy in 2012 built its own solar energy farm at Norfolk Naval Station, which can generate up to 2.1 megawatts of electricity – enough to power 200 homes.

An article in my alumni magazine told about friends of mine who retired from the local church ministry and built an eco-friendly home that will produce more energy than it uses each year. It features rooftop solar panels, 13-inch exterior walls packed with R-34 insulation and triple pane windows that are more than three inches thick. The water supply comes from 9,000 gallons of rain water in storage. If more of us did the same, the energy use of carbon fuels would diminish greatly.

And this change is happening quickly. A year ago, India was planning on building over 370 coal plants, or about 240 gigawatts of power. But just recently, the Indian government has just concluded they won't need any additional coal plants to be built for another decade, in large part because they have been pushing hard towards a goal of building 275 gigawatts of renewable power by the year 2027.
And when a company thinks of building a coal plant that's going to be around for 40 years, the economics of it are increasingly not making sense.

The Ecomii website defines carbon footprint as
"A carbon footprint is an estimate of how much carbon dioxide is produced to support your lifestyle. Essentially, it measures your impact on the climate based on how much carbon dioxide you produce. Factors that contribute to your carbon footprint include your travel methods and general home energy usage. Carbon footprints can also be applied on a larger scale, to companies, businesses, even countries.
The national average for someone in the US is 7.5 tons of carbon dioxide emitted per year. Start by calculating your own carbon footprint using an online calculator and think of ways you can modify your lifestyle to reduce it. Even better, try to offset your carbon footprint if you can! Individuals and businesses can help offset their carbon dioxide production by buying carbon offsets or planting trees to balance the carbon dioxide emissions."[149]

We have looked at creation myths from around the world. And at the first one found in Genesis. We have directed your attention to the Big Bang theory of creation ad Darwin's theory of evolution. As we ended our book, we looked to several experts to explain what is and has been happening to our earth. We just finished a review of what the United Methodist Church believes its members should do to protect the earth. Now, we will simply list some things we can all do to help curb carbon dioxide emissions.

1. Reduce emissions by:
    a. Using public transportation.
    b. Inflating tries to appropriate levels.
    c. Combining trips since cars emit more carbon dioxide when cold.
    d. Do not let cars idol to warm the interior in cold weather.
    e. Purchasing a vehicle that gets good gas mileage, a hybrid vehicle, or an electric vehicle.
    f. Removing excess weight from your vehicle.
    g. Change any filters that could help your engine get better gas mileage.
    h. Drive like a truck. One day I followed a truck on the interstate for about two hours. When it went downhill, it sped up and when it went up hill, it slowed down. I was driving a hybrid and got 56 miles to a gallon that trip.
    i. Start slowly, go slowly, and stop slowly. The jack rabbit starts consume a lot of gas.

---

[149] Ecomii web site.

2. We need to use more renewable energy.
    a. Solar power is the easiest way to produce energy. By putting a few panels on your roof, most of your electrical needs will be generated on your own roof.
    b. Wind turbines can be seen in Indiana, Pennsylvania and many other states. On a cloudy day, the wind may be blowing to generate energy.
    c. Geothermal energy is generated and stored in the Earth. Thermal energy is the energy that determines the temperature of matter. The geothermal energy of the Earth's crust originates from the original formation of the planet and from radioactive decay of materials.
    d. Wells and heat pumps accomplishes the seemingly impossible trick of warming our house with cool water through the use of a water-to-air heat pump. The cost is about 20% less that a gas or electric bill.
    e. Hydroelectric energy comes from water moving turbines in dams or rivers. The Hoover dam is an example.

3. We must increase our energy efficiency.
    a. Turn off the light.
    b. Turn down the heat.
    c. Turn up the air condition.
    d. Have your heating and air conditioning system inspected annually.
    e. Change furnace filters monthly.
    f. Do not exceed the speed limit while driving.
    g. Close vents to unused rooms.
    h. Make sure vents are not blocked by curtains or furniture.
    i. Close curtains or blinds to eliminate the need to cool a room in the summer that has the sun coming it.

4. Providing more efficient transportation.
    a. Most of the pick-up trucks and SUV's on the road today are huge. Do we really need vehicles that large?
    b. We do have electric cars. They are costly. Hopefully, as time goes by, they will become affordable for more people.
    c. Car-polling for work is still a good option. Most metropolitan areas have park-and-ride lots as commuters gather to go into the nearby city.

5. Stop deforestation and begin to plant more trees.
    a. Plant fruit and nut trees that will provide fruit and nuts to you and the animals as well as serve as a means to help clear the air of carbon dioxide.

6. The following are a few simple electricity saving activities that can help reduce carbon dioxide emissions:
    a. Purchase energy efficient appliances, electronics, and water heaters.

- b. Insulate water heaters.
- c. Properly insulate your home to reduce heating and cooling needs.
- d. Change light bulbs to energy efficient light bulbs. These last much longer than standard bulbs and provide energy savings.

7. Recycle

    We are amazed at how many cities are not providing a recycling program. Encourage yours to do so if they are not.
    - a. Recycle plastic, aluminum, paper, cardboard and glass.

    The more we can recycle, the less we have to manufacture from scratch.
    - b. Do not litter cans or glass bottles. Recycle them.
    - c. Reuse paper or plastic bags from the store for trash.

8. Buy or use canvas tote bags that you can take back to the store to reuse rather than choosing from "paper or plastic."

Further information on how you can reduce greenhouse gas emissions is located at the following websites:

Environmental Protection Agency website – What you can do to reduce emissions
http://www.epa.gov/climatechange/wycd/

Fuel Economy website – http://www.fueleconomy.gov

Bibliography

Alva-Basurto, Jorge Christian and Jesus Ernesto Arias-Gonzalez, "Modelling the Effects of Climate Change on a Caribbean Coral Reef Food Web," *Ecological Modelling* 289 (2014).

Anderson, Bernhard W., *Understanding the Old Testament*, Englewood Cliffs, New Jersey, 1966.

Berry, R.J., editor *Environmental Stewardship,* London, New York, 2006.

Biello, David, "Ocean Impact Map Reveals Human Reach Global As vast as the oceans are, almost no waters remain untouched by human activities," *Scientific American* (February 15, 2008).

Bijma, J., et al., "Climate Change and the Oceans – What Does the Future Hold?," *Marine Pollution Bulletin* 74, issue 2 (2013).

Blake, Eric S. and Ethan J. Gibney, *The Deadliest, costliest, and most intense United States tropical cyclones from 1851 to 2010 (and other frequently requested hurricane facts)*, (Miami: NOAA Technical Memorandum NWS NHC-6, 2011).

Bryant, Dirk, et al, *The Last Frontier Forests: Ecosystems and Economies on the Edge*, Washington, D.C., 1997.

Bryne, Kevin, "What Effects Can the Full Moon Have on Weather, People and Animals?" *Accuweather*.

Burke, Lauretta et al., *Reefs at Risk Revisited,* (Washington D.C., 2011).

Butler, Rhett A., "Coral Reefs Decimated by 2050, Great Barrier Reef's Coral 95% Dead." *Mongabay*.

Case, Michael J., et al., "Relative Sensitivity to Climate Change of Species in Northwestern North America," *Biological Conservation* 187 (2015).

Chiu, M.C., et al., "Climate-change Influences on the Response of Macroinvertebrate Communities to Pesticide Contamination in the Sacramento River, California Watershed," *Science of the Total Environment* 581-582 (2017).

Climate and Weather website, http://www.climateandweather.net/global-warming/climate-change-and-animals.html.

Conserve Energy Future website, http://www.conserve-energy-future.com/30-astounding-ways-to-protect-and-conserve-wildlife.php.

Conserve Energy Future website, http://www.conserve-energy-future.com/fabulous-ways-to-protect-trees-and-conserve-forests.php.

Darwin, Charles R., *Journal of Researches into the Geology and Natural History of the Various Countries Visited During the Voyage of H.M.S. 'Beagle', Uunder the Command of Captain FitzRoy, R. N. from 1832 to 1836*, London, 1842.

Darwin, Charles R., *The Origin of Species*, New York, 2004.

Dell'Amore, Christine, "7 Species Hit Hard by Climate Change—Including One That's Already Extinct," *National Geographic*, April 2, 2014.

DeNoon, Daniel J., "Sea Squirt Drug Offers Cancer Hope." *WebMD*.

Endangered.org website, http://www.endangered.org/10-easy-things-you-can-do-to-save-endangered-species/.

European Space Agency website, http://www.esa.int/Our_Activities/Space_Science/Herschel/How_many_stars_are_there_in_the_Universe.

Gascon, Claude, et al., "Deforestation and Forest Fragmentation in the Amazon. In Lessons from Amazonia," in *The Ecology and Conservation of a Fragmented Forest,* ed. Bierregaard, Richard O. Jr., et al., New Haven and London, 2001.

Gifford, Clive, *How the World Works*, London 2013.

Goldenberg, Suzanne, "West Antarctic ice collapse 'could drown Middle East and Asia crops,'" *The Guardian* (2014).

Grant, Peter R., *Ecology and Evolution of Darwin's Finches*, Princeton, New Jersey, 1986.

Karl, Thomas R., et al., eds., *Weather and Climate Extremes in a Changing Climate. Regions of Focus: North America, Hawaii, Caribbean, and U.S. Pacific Islands*, Washington, DC, 2008.

Hamilton, Adam, *Making Sense of the Bible,* New York, New York, 2014.

Hamilton, Edith, *Mythology*, New York, Boston & London, 1942.

Hamilton, Virginia and Moser, *In the Beginning*, San Diego, New York, London, 1988.

Hassol, Susan Joy, *Impacts of a Warming Arctic: Arctic Climate Impact Assessment*, Cambridge, 2004.

Hawking, Stephen A., *A Brief History of Time*, New York, 1988.

Karl, Thomas R. et al., editors, *Global Climate Change Impacts in the United States*, New York, 2009.

Klingbel, Andiens, Editor. *The Genesis Creation Account and it's Reverberations in the Old Testament.* Berrien Springs, Michigan, 2015.

Krause, Bernie and Almo Farina, "Using Ecoacoustic Methods to Survey the Impacts of Climate Change on Biodiversity," *Biological Conservation* 195 (2016).

Mays, James L., general editor. *Harper Collins Bible Commentary*, San Francisco, 1988.

McKibben, Bill, *Earth-Making a Life on a New Planet,* New York, New York, 2010.

Mac Nally, Ralph, et al., "Forecasting the Impacts of Habitat Fragmentation. Evaluation of Species-specific Predictions of the Impact of Habitat Fragmentation on Birds in the Box-ironbark Forests of Central Victoria, Australia," *Biological Conservation* 95 (2000).

msn website, http://www.msn.com/en-us/weather/topstories/january-smashed-another-global-temperature-record/ar-BBpzSlc?li=BBnb7Kz.

Mueller, Derek R., et al., "Break-up of the Largest Arctic Ice Shelf and Associated Loss of an Epishelf Lake," *Geophysical Research Letters*, 2003.

NASA website, https://climate.nasa.gov/climate_resources/24/.

NASA website, https://www.nasa.gov/topics/earth/features/thick-melt.html.

National Center for Science Education website, http://ncse.com/taking-action/ten-major-court-cases-evolution-creationism.

Neimark, Peninah and Rhodes, Peter, *The Environmental Debate, second edition,* Amenia, New York, 2011.

NOAA website, http://oceanservice.noaa.gov/facts/coral_bleach.html.

NOAA website, http://www.publicaffairs.noaa.gov/25list.html.

Olivier, Jos G.J. et al., *Trends in global CO2 emissions; 2012 Report*, The Hague, 2012.

PBS website, http://www.pbs.org/wgbh/evolution/library/01/6/l_016_02.html.

Pitcher, Tony J. and William W.L. Cheung, "Fisheries: Hope or Despair, " *Marine Pollution Bulletin* 74, issue 2 (2013).

Reed, Rebecca A., et al., "Contribution of Roads to Forest Fragmentation in the Rocky Mountains," *Conservation Biology*, 10, 4 (1996).

Rosenblatt, Kalhan, "California on Track to Have Wettest Year on Record Following Five-Year Drought," NBC News, February 23, 2017.

Sarfaty, M., et al., "Medical Alert! Climate Change Is Harming Our Health," *The Medical Society Consortium on Climate and Health.*

Science.com website, http://www.space.com/20720-earth-like-alien-planets-discovery.html.

Science Daily website, www.sciencedaily.com/releases/2011/08/110823180459.htm

SFGATE, http://www.sfgate.com/bayarea/article/Near-record-Sierra-snowpack-185-percent-of-10969482.php.

Smith, Robert L. and Thomas M. Smith, *Ecology and Field Biology*, Menlo Park, California, 2001.

Stevens, William K., "Plant Survey Reveals Many Species Threatened With Extinction," *New York Times*, April 9, 1998.

Truth in Science website, http://www.truthinscience.org.uk/site/content/view/127/65/.

United States Environmental Protection Agency, *Inventory of U.S. Greenhouse Gas Emissions and Sinks: 1990-2011*, https://www.epa.gov/sites/production/files/2015-12/documents/us-ghg-inventory-2013-main-text.pdf.

Water Use it Wisely website, http://wateruseitwisely.com/100-ways-to-conserve/.

Webster, B., "Night-time Temperatures Could Rise Above 25C Because of Climate Change," *The Times*, June 1, 2010.

Wikipedia website, https://en.wikipedia.org/wiki/List_of_Category_5_Atlantic_hurricanes.

Wikipedia website, https://en.wikipedia.org/wiki/List_of_Category_5_Pacific_hurricanes.

Williams, Michael E., editor, *The Story Teller's Companion to the Bible*. Nashville, TN. Volume I, 1976.

Woodcock, James, et al., "Public Health Benefits of Strategies to Reduce Greenhouse-gas Emissions: Urban Land Transport," *Lancet*, 2009.

World Meteorological Organization, *The Global Climate 2001-2010: A Decade of Climate Extremes - Summary Report*, Geneva, 2013.

www.ingramcontent.com/pod-product-compliance
Lightning Source LLC
Chambersburg PA
CBHW051420070526
44584CB00023B/3509